Graph theory
and combinatorics

R J Wilson
The Open University

Graph theory and combinatorics

Pitman Advanced Publishing Program
LONDON · SAN FRANCISCO · MELBOURNE

FEARON PITMAN PUBLISHERS INC.
6 Davis Drive, Belmont, California 94002

PITMAN PUBLISHING LIMITED
39 Parker Street, London WC2B 5PB
North American Editorial Office
1020 Plain Street, Marshfield, Massachusetts 02050

Associated Companies
Copp Clark Pitman, Toronto
Pitman Publishing New Zealand Ltd, Wellington
Pitman Publishing Pty Ltd, Melbourne

©R J Wilson 1979

AMS Subject Classifications: 05-xx

All rights reserved. No part of this publication may be reproduced, stored in a retrieval system, or transmitted in any form or by any means, electronic, mechanical, photocopying, recording and/or otherwise without the prior written permission of the publishers. The paperback edition of this book may not be lent, resold, hired out or otherwise disposed of by way of trade in any form of binding or cover other than that in which it is published, without the prior consent of the publishers.

Manufactured in Great Britain

ISBN 0 273 08435 6

This book is dedicated to the
memory of Derek Waller

Preface

This book presents the proceedings of a one-day conference in Combinatorics and Graph Theory held at The Open University, England, on 12 May 1978. The first nine talks presented here were given at the conference, and cover a wide variety of topics ranging from topological graph theory and block designs to latin rectangles and polymer chemistry. The tenth author, Christopher Wright, had been invited to present his talk on traffic-flow problems, but was unable to do so due to other commitments; he has kindly allowed us to publish the talk he would have given. In all cases, the authors were chosen for their ability to combine interesting expository material in the areas concerned with an account of recent research and new results in these areas.

One of the special features of the conference was a Poster Session, and one of the presentations at this session appears here. I should like to thank my colleagues John Mason and Roger Duke for organizing the Poster Session, and to thank them and Roy Nelson for helping with the academic side of the programme. On the administrative side, I should like to express my thanks to everyone involved, and in particular to Marion Aldred, Jennifer Goldrei, Joan Street and Mike Bandle. Most of all, I should like to thank Frances Thomas for her excellent typing of the entire manuscript.

Finally, this volume is dedicated to the memory of Derek Waller, who was prevented by illness from attending the conference. Three weeks later he died of leukaemia, leaving a wife and three small children. His untimely death is a sad loss for British Combinatorics.

The Open University Robin J. Wilson
January 1979

Contents

1. L.D. ANDERSEN and A.J.W. HILTON
   Generalized latin rectangles   1

2. J.-C. BERMOND
   Graceful graphs, radio antennae and French windmills   18

3. P.J. CAMERON
   Multiple transitivity in graphs   38

4. M. GORDON and J.A. TORKINGTON
   Improvements in the random walk model of polymer chains   49

5. Michel LAS VERGNAS
   On Eulerian partitions of graphs   62

6. Colin McDIARMID
   Colouring random graphs badly   76

7. C.St.J.A. NASH-WILLIAMS
   Acyclic detachments of graphs   87

8. F. PIPER
   Unitary block designs   98

9. A.T. WHITE
   Strongly symmetric maps   106

10. C.C. WRIGHT
    Arcs and cars: an approach to road traffic management based on graph theory   133

Poster session paper
    M. GORDON and J.W. KENNEDY
    Some problems on lattice graphs   147

L D Andersen and A J W Hilton
Generalized latin rectangles

1. INTRODUCTION

A <u>latin square</u> of size n × n based on the symbols 1, ..., n is an n × n matrix in which each cell is filled by exactly one symbol in such a way that each symbol occurs exactly once in each row and exactly once in each column. Instead of the n positive integers 1, ..., n any set of n symbols can be used.

A latin square can be thought of as a finite quasigroup: a <u>quasigroup</u> (Q, *) is a set Q with a binary operation * such that the equations a*x = b and y*a = b are uniquely solvable for each pair (a,b) of elements of Q. Thus the multiplication table of a quasigroup of order n is precisely a latin square of size n × n (with a headline and a sideline). Figure 1 gives an example.

Q,*	1	2	3	4
1	2	4	1	3
2	4	3	2	1
3	3	1	4	2
4	1	2	3	4

example
1*4 = 3
4*1 = 1

Latin square

Figure 1

Latin squares and quasigroups are extensively treated in [3], which contains several references also to works concerned with the application of latin squares in the design of experiments (the classic in this field is [7]).

One intriguing combinatorial problem concerning latin squares is that of embedding. There are no end of variations on this theme. We mention here two results.

H. J. Ryser [10] proved that if A is an r × s matrix where each cell contains one of the symbols 1, ..., n such that no symbol occurs more than once in any row or column, then A can be embedded in (found as a submatrix of) some latin square of size n × n on symbols 1, ..., n if and only if each symbol occurs at least r + s - n times in A.

A. Cruse [2] proved that if A is an r × r matrix where each cell contains one of the symbols 1, ..., n such that no symbol occurs more than once in any row or column and such that if cell (i,j) is occupied by the symbol k then so is cell(j,i), then A can be embedded in a symmetric latin square of size n × n on symbols 1, ..., n if and only if each symbol occurs at least 2r - n times in A and at least r different symbols occur a number of times congruent to n modulo 2. Other embedding theorems for quasigroups can be found in [9].

This paper is concerned with the following generalization of latin squares:

A (p, q, x)-latin rectangle of size r × s on symbols 1, ..., n is an r × s matrix in which each cell is filled by exactly x symbols in such a way that each symbol occurs at most p times in each row and at most q times in each column (p, q, x, r, s and n are positive integers).

A (2,2,2)-latin rectangle A (2,2,3)-latin square
of size 2x3 on symbols   of size 2x2 on symbols
1, 2, 3, 4.   1, 2, 3, 4, 5.

Figure 2

A (p, q, x)-latin rectangle in which each symbol occurs exactly p times in each row and exactly q times in each column is called <u>exact</u>.

The x symbols in a cell of a (p, q, x)-latin rectangle need not be distinct; if they are distinct for each cell the rectangle is said to be <u>without repetition</u>. Note also that a (p, p, x)-latin square may be <u>symmetric</u>.

1	3	1	3
2	4	2	4
3	1	3	1
4	2	4	2

2	1	1
3	2	3
1	3	1
2	3	2
1	1	2
3	2	3

An exact (2,1,2)-latin rectangle A on symbols 1, 2, 3, 4 without repetition

An exact symmetric (2,2,2)-latin square B on symbols 1, 2, 3, 4.

Figure 3

A (p, p, 1)-square is a <u>frequency square</u> (F-square) with frequency vector (p, p, ..., p). F-squares are treated in [8], to which we refer the reader for further references. Here we investigate (p, q, x)-latin rectangles from other points of view.

In Section 5 we generalize the theorems of Ryser and Cruse by giving necessary and sufficient conditions for the embedding of (p, q, x)-latin rectangles, with or without repetition, symmetric or possibly unsymmetric.

Sections 3 and 4 contain construction and decomposition theorems, and we apply the results and methods to questions on quasigroups and on equitable edge-colourings of graphs, defined in the next section.

Many of the results stated here have quite long proofs. Most of the theorems given (but not all) can be found with proofs in [1] and will be published elsewhere.

We wish to thank Professor Trevor Evans for a helpful discussion.

2. EQUITABLE AND BALANCED EDGE-COLOURINGS OF GRAPHS

Equitable edge-colourings of graphs relate to the work of this paper in two ways: a result about them is used in several of our proofs, and on the other hand our results can be stated as results about such colourings.

An <u>edge-colouring</u> of a graph G with colours 1, ..., k is a partition of the edges and loops of G into k mutually disjoint subsets $c_1, \ldots, c_k$ (note that any partition will do, so that an edge-colouring need not be <u>proper</u> – that is, having no colour present more than once at any vertex). An edge or loop has colour i if it belongs to c_i.

Given an edge-colouring, we let $c_i(v)$ be the set of edges and loops on vertex v of colour i, and $c_i(u,v)$ be the set of edges joining vertices u and v (the set of loops on u, if u = v) of colour i.

An edge-colouring is <u>equitable</u> if, for all vertices v,

(a) $\max_{i,j} \left| |c_i(v)| - |c_j(v)| \right| \leq 1.$

It is <u>balanced</u> if it is equitable and for each pair of vertices u, v:

(b) $\max_{i,j} \left| |c_i(u,v)| - |c_j(u,v)| \right| \leq 1.$

Thus an edge-colouring is equitable if the colours occur as uniformly as possible at each vertex, and it is balanced if in addition the colours are shared as evenly as possible on each multiple edge.

An exact (p, q, x)-latin rectangle A corresponds to an equitable edge-colouring of a bipartite graph with a vertex for each row and a vertex for each column, where each row vertex ρ_i (corresponding to the i-th row) is joined to each column vertex γ_j (corresponding to the j-th column) by x edges being coloured with the symbols in the (i,j)-th cell of A, where the number of edges of any given colour joining ρ_i to γ_j is equal to the number of occurrences of the corresponding symbol in the cell (i,j). For example, from the rectangle A of Figure 3 we get the graph of Figure 4 with edge-colouring indicated. Since A is without repetition, the colouring is balanced.

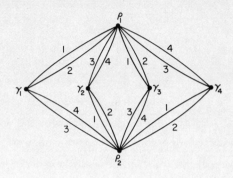

Figure 4

In the same way an exact symmetric (p, p, x)-latin square corresponds to an equitable edge-colouring of a complete graph with loops and multiple edges. We shall not go into details, but just remark that several of the following results can be seen in this light, although we only state one such result (Theorem 9 in Section 4).

The notion of equitable and balanced edge-colourings was introduced by D. de Werra, and he also proved the following important result ([4] - [6]). We give a short proof of our own.

Theorem 1. For each $k \geq 1$, any bipartite graph has a balanced edge-colouring with k colours.
Proof. Colour the edges of the graph in any way such that (b) is satisfied. The condition only affects each multiple edge by itself, so this is clearly possible. We then modify the colouring to make (a) be fulfilled without violating (b). Suppose that at some vertex v,

$$\max_{i,j} \left| |c_i(v)| - |c_j(v)| \right| > 1,$$

and suppose that the maximum is attained for colours 1 and 2. We can

assume that $|c_1(v)| > |c_2(v)| + 1$. Let P be a maximal chain
$v = v_0, v_1, v_2, \ldots, v_h$ such that the edges are coloured alternately 1 and 2
(the edge joining v_0 and v_1 having colour 1), and such that
$|c_1(v_i, v_{i+1})| = |c_2(v_i, v_{i+1})| + 1$ when i is even,
$|c_2(v_i, v_{i+1})| = |c_1(v_i, v_{i+1})| + 1$ when i is odd, and P uses only one edge
from each multiple edge. Now $h \neq 0$, because v has some neighbour v_1 for
which $|c_1(v, v_1)| = |c_2(v, v_1)| + 1$, since $|c_1(v)| > |c_2(v)| + 1$. Also
$v_h \neq v_0$, because if $v_j = v_0$ then j is even (the graph is bipartite), i.e.
both colours occur the same number of times in total on the multiple edges
incident with v_0 used so far, so the chain can be continued because
$|c_1(v)| > |c_2(v)| + 1$. Interchanging the two colours 1 and 2 on the chain
P clearly does not violate (b), and it reduces the number of pairs of
colours for which $||c_i(v)| - |c_j(v)||$ was maximal (greater than 1) by at
least 1. We only have to check that

$$\max_{i,j} \left| |c_i(v_h)| - |c_j(v_h)| \right|$$

is not increased. Repeated application of the argument then proves the
theorem. But if h is even the maximality of P implies that the colour 2
occurs at least once more than the colour 1 at v_h which implies the
required result, and a similar argument holds if h is odd. This proves
Theorem 1. □

It is not true in general that any graph has an equitable edge-colouring
with k colours for all k.

3. EXISTENCE AND CONSTRUCTION OF EXACT (p, q, x)-LATIN RECTANGLES

An exact (p, q, x)-latin rectangle on xt symbols has pxt occurrences in
each row, so a row must contain pt cells; similarly a column contains qt
cells, so the rectangle has size qt × pt.

Theorem 2. Let p, q, x be positive integers and let t be a rational
number such that pt, qt, xt are positive integers. Then there exists an
exact (p, q, x)-latin rectangle on xt symbols and an exact symmetric
(p, p, x)-latin square on xt symbols. If $t \geq 1$, both can be taken to be
without repetition. □

A simple method to obtain exact (p, q, x)-latin rectangles which are symmetric if p = q and without repetition if t ≥ 1 is the cyclic way of placing the symbols exemplified in Figure 5. The rectangle A of Figure 3 is also of this type.

1	4	7	3	6	2	5	
2 3	5 6	1 2	4 5	7 1	3 4	6 7	p = 3
4	7	3	6	2	5	1	x = 3
5 6	1 2	4 5	7 1	3 4	6 7	2 3	$t = \frac{7}{3}$
7 1							

Figure 5

There are several ways of getting generalized latin rectangles from others. Some are rather trivial, like placing several rectangles on top of or next to each other. The following is more interesting, yet simple. Figure 6 provides an example.

1 2 3	4 5 6	1 3 6	2 4 5
4 5 6	1 2 3	2 4 5	1 3 6

An exact (2,1,3)-latin rectangle A on 1, 2, 3, 4, 5, 6.

↓ Identify 1 and 4, 2 and 5, 3 and 6.

1 2 3	1 2 3	1 3 3	2 1 2
1 2 3	1 2 3	2 1 2	1 3 3

An exact (4,2,3)-latin rectangle B on 1, 2, 3.

Figure 6

In brief, let m, n be positive integers and A an exact (p, q, x)-latin rectangle on 1, ..., mn. Identify symbols i and j whenever $i \equiv j \pmod{m}$. Then an exact (np, nq, x)-latin rectangle B on 1, ..., m is obtained. B is called the <u>modulo m reduction</u> of A. So in Figure 6, B is the modulo 3 reduction of A.

This process has a converse as stated in the next theorem.

<u>Theorem 3</u>. Let B be an exact (np, nq, x)-latin rectangle on 1, ..., m. Then
(i) B is the modulo m reduction of some exact (p, q, x)-latin rectangle A on 1, ..., mn;
(ii) if no symbol occurs more than n times in any cell of B then A can be taken to be without repetition. ☐

Since we go from B to A by splitting each symbol into n symbols, (ii) of Theorem 3 is best possible.

The theorem does not hold for symmetric squares - that is, we cannot be sure that A can be chosen to be a symmetric square even if B is (the square B of Figure 3 is a counterexample); we have only the following:

<u>Theorem 4</u>. Let B be an exact symmetric (2np, 2np, x)-latin square on symbols 1, ..., m. Then B is the modulo m reduction of some exact symmetric (2p, 2p, x)-latin square A on 1, ..., mn. ☐

In the next section we consider another way of obtaining (p, q, x)-latin rectangles from others.

4. MERGING OF ADJACENT CELLS

1	2	1	3	1	2	2	3	3
2	2	3	1	2	1	3	3	1
2	3	3	1	1	2	3	2	1
1	3	2	2	2	3	1	1	3
3	1	1	3	3	1	2	2	2
3	1	2	2	3	3	1	1	2

An exact $(3,2,1)$-latin rectangle A of size 6×9

) erase lines in block

1	2	1	3	1	2	2	3	3
2	2	3	1	2	1	3	3	1
2	3	3	1	1	2	3	2	1
1	3	2	2	2	3	1	1	3
3	1	1	3	3	1	2	2	2
3	1	2	2	3	3	1	1	2

An exact $(6,6,6)$-latin square B of size 3×3

Figure 7

Figure 7 gives an example of the following process:
Let A be an exact (p, q, x)-latin rectangle of size mr × ns for positive integers m, n, r, s. Merge (identify) cells (i_1, j_1) and (i_2, j_2) whenever $\lceil i_1/m \rceil = \lceil i_2/m \rceil$ and $\lceil j_1/n \rceil = \lceil j_2/n \rceil$ (where $\lceil z \rceil$ denotes the least integer not smaller than z). Then an exact (mp, nq, mnx)-latin rectangle B of size r × s is obtained. B is called the (m, n)-merger of A. For example, in Figure 7, B is the (2, 3)-merger of A.

In the following, we shall use the term "merging of cells" about a slightly more general situation as well.

9

We have a converse theorem:

Theorem 5. Let B be an exact (mp, nq, mnx)-latin rectangle. Then
(i) B is the (m, n)-merger of some exact (p, q, x)-latin rectangle A;
(ii) if no symbol occurs more than mn times in any cell of B then A can be taken to be without repetition. □

To go from B to A we subdivide each cell of B into mn cells, so (ii) is best possible.

We illustrate the proof of Theorem 5 by proving a related theorem about quasigroups. Let $(Q,*)$ be a quasigroup. Suppose we have two partitions of the set Q, $Q = A_1 \cup \ldots \cup A_r = B_1 \cup \ldots \cup B_s$, the A_i's being mutually disjoint, likewise the B_j's. Form an $r \times s$ matrix M, where cell (i, j) contains each $a*b$ with $a \in A_i$, $b \in B_j$, counting repetitions. Then cell (i, j) contains $|A_i| \cdot |B_j|$ elements. We call M the <u>set-multiplication table</u> for $(Q,*)$ corresponding to the two partitions. It is easy to see that row i of M contains each element of Q $|A_i|$ times, and column j each element $|B_j|$ times.

The next theorem shows that if we write down any table with these properties, then it is the set-multiplication table for some pair of partitions of some quasigroup. In other words, we can postulate any set-multiplication table and be sure that there is a quasigroup satisfying it.

We define an SM-matrix on σ elements to be any matrix (say $r \times s$) where there are positive integers $\rho(i)$ associated with each row and $\gamma(j)$ with each column such that

$$\sigma = \sum_{i=1}^{r} \rho(i) = \sum_{j=1}^{s} \gamma(j),$$

each cell (i, j) contains $\rho(i)\gamma(j)$ elements, and each element occurs $\rho(i)$ times in row i and $\gamma(j)$ times in column j.

Theorem 6. Any SM-matrix on σ elements is the set-multiplication table for some quasigroup on these elements.

<u>Proof.</u> Let A be an SM-matrix of size $r \times s$ on elements $\alpha_1, \ldots, \alpha_\sigma$. We show that if $r \neq \sigma$ then A can be obtained from an $(r + 1) \times s$ SM-matrix by merging the cells of two rows (such that any pair of cells in the same column are identified). Repeated application of this argument first on the

rows, and then on the columns, shows that A can be obtained by "generalized merging" from a $\sigma \times \sigma$ SM-matrix B. But B is just a latin square, a quasigroup, and A is easily seen to be a set-multiplication table for it.

If $r \neq \sigma$ then some $\rho(i)$ is at least 2. Assume without loss of generality that $\rho(1) \geq 2$. We want to split the first row of A into new rows. Construct a bipartite graph G with vertex classes $\{v_1, \ldots, v_s\}$ and $\{\alpha_1, \ldots, \alpha_\sigma\}$, where v_i is joined to α_j by k edges if and only if the symbol α_j occurs k times in cell (1,i). Then v_i has degree $\rho(1)\gamma(i)$ and α_j has degree $\rho(1)$. Give G an equitable edge-colouring with $\rho(1)$ colours. Let c_1 be some colour class. Then each v_i has exactly $\gamma(i)$ edges of colour 1 on it, and each α_i is on exactly one such edge. Split row 1 of A into two rows 1' and 1" where a symbol α_j goes in row 1' in cell (1', i) if and only if there is an edge of colour 1 joining v_i and α_j, and symbol α_j goes in row 1" in cell (1", i) as many times as there are edges joining v_i and α_j of colour different from 1. Let $\rho(1') = 1$, $\rho(1") = \rho(1) - 1$. It is easy to see that we have obtained an SM-matrix of size $(r + 1) \times s$. This proves Theorem 6. □

Corresponding to (i) of Theorem 5 we have the following result about symmetric squares.

Theorem 7. Let B be an exact symmetric (mp, mp, m^2x)-latin square. Then B is the (m, m)-merger of some exact symmetric (p, p, x)-latin square A if and only if at most mx distinct symbols occur an odd number of times in any give diagonal cell of B. □

The condition is due to the fact that a symbol occurring an odd number of times in a diagonal cell of B must occur in one of the corresponding diagona cells of A. If we want to know when A can be taken to be without repetition we get a rather complicated necessary and sufficient condition; we omit the result here.

We have the following related result about quasigroups.

Theorem 8. Any symmetric SM-matrix on σ elements with at most $\rho(i)$ distinct elements occurring an odd number of times in the i-th diagonal cell is the set-multiplication table for some commutative quasigroup on these elements.□

We state here one of the more important results about edge-colourings which follow from the previous theorems. Let H_n^e be a graph on n vertices where each pair of vertices are joined by e edges and there are e loops on each vertex. When identifying (merging) two vertices, each edge joining them gives rise to two loops.

Theorem 9. Let m, x, xt, p, pt, pt/m be positive integers. Any equitable edge-colouring of $H_{pt/m}^{m^2x}$ with xt colours, such that at any given vertex not more than mx distinct colours occur on an odd number of loops, can be obtained by merging sets of m vertices of some H_{pt}^x which is equitably coloured with xt colours. □

5. EMBEDDING THEOREMS

In this section we consider embeddings of (p, q, x)-latin rectangles and obtain results like the two by Ryser and by Cruse mentioned in the Introduction. We need some notation:

If A is a generalized latin rectangle, let

$N_A(i)$ be the number of times symbol i occurs in A,

$N_\rho(i)$ the number of times i occurs in row ρ of A,

$N_\gamma(i)$ the number of times i occurs in column γ of A.

Let p, q, x, r, s, pt, qt, xt be positive integers.

Theorem 10. A (p, q, x)-latin rectangle A of size r × s on 1, ..., xt can be embedded in an exact (p, q, x)-latin rectangle on the same symbols if and only if, for all i,

$$N_A(i) \geq qs + pr - pqt. \quad \square$$

Theorem 11. A (p, q, x)-latin rectangle A without repetition of size r × s on 1, ..., xt can be embedded in an exact (p, q, x)-latin rectangle without repetition on the same symbols if and only if, for all i,

$$N_A(i) \geq qs + pr - pqt,$$

$$N_A(i) \leq qs + pr - pqt + (pt-s)(qt-r),$$

$$N_\rho(i) \geq p - (pt-s) \quad \text{for all rows } \rho \text{ of } A,$$

$$N_\gamma(i) \geq q - (qt-r) \quad \text{for all columns } \gamma \text{ of } A. \quad \square$$

The necessity of the conditions of Theorems 10 and 11 is not difficult to prove. Suppose A is embedded in an exact (p, q, x)-latin rectangle as on Figure 8.

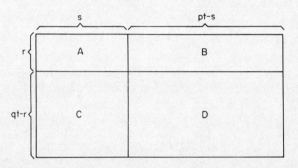

Figure 8

Then $N_{A \cup B}(i) = pr$ for each symbol i, and $N_{B \cup D}(i) = q(pt-s)$. Now
$N_A(i) = N_{A \cup B}(i) - N_B(i) \geq pr - N_{B \cup D}(i) = pr - q(pt-s),$
which proves the necessity of the condition of Theorem 10. If the exact rectangle is without repetition, $N_D(i) \leq (qt-r)(pt-s)$, and hence

$$N_A(i) = N_{A \cup B}(i) - N_B(i) = N_{A \cup B}(i) - (N_{B \cup D}(i) - N_D(i))$$
$$\leq pr - q(pt-s) + (qt-r)(pt-s),$$

proving the necessity of the second condition of Theorem 11. The two last conditions have even simpler explanations.

Theorem 12. A symmetric (p, p, x)-latin square A of size $r \times r$ on symbols $1, \ldots, xt$ can be embedded in an exact symmetric (p, p, x)-latin square on the same symbols if and only if

$N_A(i) \geq 2pr - p^2t$ for all symbols i,

and $N_A(i) \not\equiv p^2t \pmod{2}$ for at most $x(pt-r)$ distinct symbols i.   □

The last condition is due to the fact that each symbol occurs exactly p^2t times in an exact (p, p, x)-latin square. A symbol i for which $N_A(i) \not\equiv p^2t \pmod{2}$ must occur on the remaining diagonal which has $x(pt-r)$ available places.

We finally present the necessary and sufficient condition for the embedding of symmetric squares without repetition. In this case there are additional difficulties. Consider the symmetric $(2, 2, 2)$-latin square A on 1, 2, 3, 4, 5, 6 in Figure 9.

1 2	1 6	2 6	3 4		
1 6	3 4	3 5	2 6		
2 6	3 5	5 6	1 4		
3 4	2 6	1 4	5 6		

Figure 9

14

An exact (2, 2, 2)-latin square on 6 symbols has size 6 × 6. The square above satisfies all conditions of Theorems 11 and 12, so it can be embedded without repetition if we do not require symmetry, and it can be embedded in an exact symmetric square if we allow repetition. It cannot, however, be embedded in an exact <u>symmetric</u> (2, 2, 2)-latin square on 6 symbols <u>without repetition</u>. This can be seen in the following way: each of the symbols 1, 2, 3, 4 occurs an odd number of times in A; since they must occur an even number of times in the exact square, each must occur in one of the remaining diagonal cells. The symbol 6 occurs twice in each column of A, so to appear twice in each of the last two rows of the exact square it must occur in the cells also in the two last columns, which requires that it occurs in each of the remaining diagonal cells. But it is impossible to have 1, 2, 3, 4, 6, 6 in the four places there.

This example partly explains the complicated condition in the theorem below.

<u>Theorem 13</u>. Let A be a symmetric (p, p, x)-latin square of size r × r without repetition on symbols 1, ..., xt. For each j, $0 \leq j \leq pt - r - 1$, let M_j and m_j be the following sets of symbols:

$$M_j = \{i | N_A(i) = 2pr - p^2t + (pt-r)^2 - j\},$$

and $m_j = \{i | N_A(i) = 2pr - p^2t + j\}.$

Furthermore, let

$$W = \{i | N_A(i) \leq 2pr - p^2t + (pt-r)^2 - (pt-r), \text{ and } N_A(i) \not\equiv p^2t \pmod{2}\},$$

and $w = \{i | N_A(i) \geq 2pr - p^2t + (pt-r), \text{ and } N_A(i) \not\equiv p^2t - (pt-r) \pmod{2}\}.$

Then A can be embedded in an exact symmetric (p, p, x)-latin square without repetition on the same symbols if and only if

$$2pr - p^2t \leq N_A(i) \leq 2pr - p^2t + (pt-r)^2 \text{ for all symbols } i,$$

$N_\rho(i) \geq p + r - pt$ for all rows ρ of A and for all symbols i,

$$\sum_{j=0}^{pt-r-1} (pt-r-j)|M_j| + |W| \leq x(pt-r) \leq xt(pt-r) - \sum_{j=0}^{pt-r-1} (pt-r-j)|m_j| - |w|.$$

6. FURTHER TOPICS

We refer to [1] for a further discussion on generalized latin rectangles. Topics treated there but not in the present paper include: partial (p, q, x)-latin rectangles (where some cells may contain fewer than x elements) and embeddings of them; symmetric (p, p, x)-latin squares with empty diagonal, corresponding to edge-colourings of graphs without loops; filling in (p, q, x)-latin rectangles symbol by symbol; the complement of a (p, q, x)-latin rectangle without repetition; and the direct product and the singular direct product of generalized latin rectangles.

REFERENCES

1. L. D. Andersen and A. J. W. Hilton, On Constructing and Embedding Generalized Latin Rectangles, Preprint, Dept. of Mathematics, University of Reading.

2. A. Cruse, On embedding incomplete symmetric latin squares, J. Combinatorial Theory (A) 16 (1974), 18-22; MR 48-8265.

3. J. Dénes and A. D. Keedwell, Latin Squares and their Applications, Academic Press, New York-London, 1974; MR 50-4338.

4. D. de Werra, Balanced schedules, INFOR 9 (1971), 230-237; MR 45-1685.

5. D. de Werra, A few remarks on chromatic scheduling, Combinatorial Programming: Methods and Applications (ed. B. Roy), D. Reidel Publ. Co., Dordrecht, Holland (1975), 337-342.

6. D. de Werra, On a particular conference scheduling problem, INFOR 13 (1975), 308-315.

7. R. A. Fisher, The Design of Experiments, Oliver and Boyd, Edinburgh, 1935.

8. A. Hedayat and E. Seiden, F-square and orthogonal F-square design: A generalization of latin square and orthogonal latin squares design, Proc. Second Chapel Hill Conf. on Combinatorial Mathematics and its Applications, Chapel Hill (1970), 261-275; MR 42-2956.

9. C. C. Lindner and T. Evans, Finite Embedding Theorems for Partial Designs and Algebras, Collection Séminaire de Mathématiques Supérieures 56, Les Presses de l'Université de Montréal, Montreal, 1977.

10. H. J. Ryser, A combinatorial theorem with an application to latin rectangles, Proc. Amer. Math. Soc. 2 (1951), 550-552; MR 13-98.

L. D. Andersen and A. J. W. Hilton
Department of Mathematics
The University of Reading
Reading
Berks RG6 2AX
England

J -C Bermond
Graceful graphs, radio antennae and French windmills

A graph G with e edges can be labelled by assigning values to its vertices and assigning to each edge the absolute value of the difference of the numbers at its endpoints. G is said to be graceful if its vertex-labels are distinct members of the set $\{0,1, \ldots, e\}$, and its edge-labels form exactly the set $\{1,2, \ldots, e\}$.

In this paper we survey the theory of graceful graphs and apply it to a problem on the configurations of antennae in radio-astronomy; this problem is related to graceful numberings of some windmills.

1. INTRODUCTION

A graph $G = (X,E)$ is <u>numbered</u> if each vertex v is assigned a non-negative integer $\phi(v)$, and each edge $\{v,w\}$ is assigned the absolute value of the difference of the numbers at its endpoints - that is, $|\phi(v) - \phi(w)|$.

The numbering is called <u>graceful</u> if, furthermore, we have:
(a) the vertices are labelled with distinct integers (that is, ϕ is an injection);
(b) the largest value of the vertex-labels is equal to the number of edges (that is, $\max_v \phi(v) = |E| = e$);
(c) all the edges of G have distinct labels chosen from the set $\{1,2, \ldots, e\}$.

A graph which admits a graceful numbering is said to be a <u>graceful graph</u>. For example, let K_n denote the complete graph on n vertices and C_n denote the cycle of length n. Then Figure 1 gives graceful numberings of K_3, K_4, $K_5 - e$, and C_4.

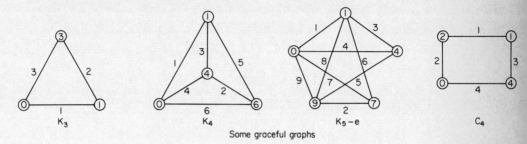

Some graceful graphs

Figure 1

Not all graphs are graceful. We leave it to the reader to check that the graphs of Figure 2 (that is, $K_3 \cup K_2$, K_5, C_5, and the graph consisting of two K_4's with a vertex in common) are not graceful; see later for a proof that K_5 and C_5 are not graceful, and [2] for the last graph.

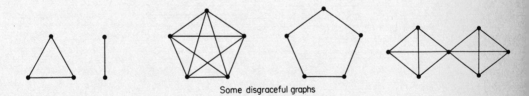

Some disgraceful graphs

Figure 2

The term 'graceful graph' (or 'graceful numbering') is recent and due to S. W. Golomb [15], and has been popularised by the articles of M. Gardner [12] and S. W. Golomb [16]. But the study of graceful graphs began much earlier. In graph theory the interest in this topic arose from a conjecture of G. Ringel [33] and an article of A. Rosa [34]. (A graceful numbering of a graph G was called a β-valuation by Rosa; note that in Rosa's article other numberings and valuations are also considered.)

This topic has also been considered in other parts of mathematics, in particular in additive number theory under the name of 'restricted difference basis'. More precisely, the problem is as follows: a <u>restricted difference basis</u> (with respect to an integer e) is a set of integers $0 = a_1 < a_2 < \ldots < a_{n-1} < a_n = e$, such that every positive integer m with $1 \leq m \leq e$ can be represented in the form $m = a_j - a_i$. The problem is then to find, for a given e, the minimum value n(e) of n, the number of integers in such a restricted difference basis. A more tangible way of posing this question is in terms of the ruler problem: to construct a ruler measuring all integral distances from 1 to the length e of the ruler by marking it at the minimum number of places. Rulers for e = 6 and 9 are shown in Figure 3.

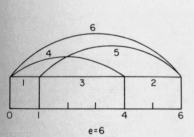 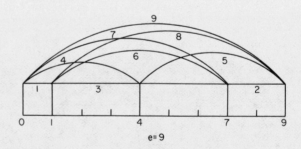

Figure 3

These correspond respectively to the graceful numberings of K_4 and $K_5 - e$, shown in Figure 1, the distances marked off being the vertex-labels. An equivalent problem is to determine the minimum number of vertices in a graceful graph, where the number of edges is prescribed. The first papers on this problem seem to be those of A. Brauer (1945), P. Erdös and I. S. Gál (1948) and L. Rédei and A. Rényi (1948) (see the article of J. Leech [29]). A survey on this problem has been written by J. C. P. Miller [30]. We shall return to this problem in Section 4.

Finally, this topic has also arisen in many applications (for example, coding theory, X-ray crystallography, radar, communication networks and astronomy). We omit further details of these applications, since they are discussed, together with numerous references, in an excellent survey paper by G. S. Bloom and S. W. Golomb [6] (see also [7]). The main unsolved problem is to determine which graphs are graceful. This appears to be very difficult, one reason being that a subgraph of a graceful graph need not be graceful (for example, C_5 is not graceful, although it is a subgraph of K_5). On the other hand, the answer is known for certain classes of graphs. Our object here is to describe the current state of knowledge; since some results has been rediscovered many times, we will mention only those authors we believe to have originated the various results.

2. SOME CLASSES OF GRACEFUL GRAPHS

We begin with a necessary condition due to Rosa [34].

Lemma 2.1. If G is a graceful Eulerian graph with e edges, then $e \equiv 0$ or 3 (mod 4).

Proof Since G is Eulerian, $\sum |\phi(v) - \phi(w)|$ ($vw \in E$) is an even number. But since G is graceful, this number is equal to the sum of the valuations of the edges, that is

$$\sum_{i=1}^{e} i = \tfrac{1}{2} e(e + 1).$$

Thus $e \equiv 0$ or 3 (mod 4). □

For example, K_5 and C_5 are Eulerian, but they have 10 and 5 edges, and thus by Lemma 2.1 they are not graceful. For cycles the necessary condition in Lemma 2.1 has been proved (by Rosa [34]) to be sufficient:

<u>Theorem 2.2</u>. The cycle C_n is graceful if and only if $n \equiv 0$ or $3 \pmod 4$. ☐

Rosa also proved the following result:

<u>Theorem 2.3</u>. All complete bipartite graphs are graceful. ☐

In Figure 4 we give a graceful numbering of $K_{m,n}$.

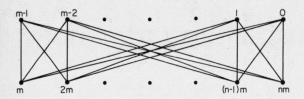

Figure 4

C. Hoede and H. Kuiper [22], and R. Frucht have proved:

<u>Theorem 2.4</u>. All wheels are graceful. ☐

Many problems remain open, however - in particular, the following conjecture due to H. Bodendiek, H. Schumacher and H. Wegner, who have proved it in certain special cases [9]:

<u>Conjecture</u>. The graphs consisting of a cycle plus one edge joining two non-adjacent vertices are graceful.

In [9] they proved also that the prism P_k consisting of two cycles $v_1, v_2, \ldots, v_k$ and $v_1', v_2', \ldots, v_k'$ plus the edges $v_i v_i'$ for $i = 1, 2, \ldots, k$ (this graph has 2k vertices and 3k edges) is graceful if $k \equiv 0 \pmod 4$. Recently R. Frucht has proved:

Theorem 2.5. All prisms are graceful. □

(Figure 5 shows a graceful numbering of the prism P_5.)

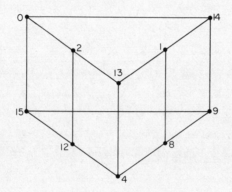

Figure 5

R. Frucht also proved that 'crowns' (cycles plus edges pendant at each vertex) are graceful. T. Gangopadyay and S. P. Rao Hebbare proved some related classes of graphs to be graceful, in particular that the prisms P_k are graceful for k even. They also conjectured that the k-cube is graceful: with A. Germa and M Makeo we have proved that this is true. To conclude this section we note the following result of D. A. Sheppard [36]:

Theorem 2.6. There are exactly e! gracefully number graphs with e edges. □

(Two graceful graphs are considered as different if either they are not isomorphic, or if they are isomorphic but have distinct graceful numberings.)

3. GRACEFUL TREES

Trees are certainly the most fascinating graphs in this area. A. Kotzig (see Rosa's article [34]) has made the following conjecture:

Conjecture. All trees are graceful.

Despite many efforts, this conjecture has not been proved; however, many classes of trees have been shown to be graceful, and a wide literature exists on the subject. A survey on this conjecture and on 'the continuing quest to call all trees graceful' has been given by G. S. Bloom [5].

The conjecture is often attributed to G. Ringel [33], but he gave only the following weaker conjecture:

Conjecture. Let T be a given tree with n vertices and n - 1 edges; then the edges of K_{2n-1} can be partitioned into 2n - 1 trees isomorphic to T.

The fact that the second conjecture will imply the first one has been shown by Rosa [34]. To see this, let the vertices of T be $v_1, \ldots, v_i, \ldots, v_n$, and let ϕ be a graceful numbering; then if the vertices of K_{2n-1} are labelled with the integers $0, 1, \ldots, 2n-2$, the 2n-1 trees T_j of the decomposition ($0 \leq j \leq 2n - 2$) are isomorphic to T and have respectively as vertex v_i the vertex of K_{2n-1} labelled $\phi(v_i) + j$ (where this number is to be taken modulo 2n - 1). In this case, the decomposition is said to be <u>cyclical</u>. More generally, Rosa [34] has shown, exactly as above, the following result:

Theorem 3.1. If G is a graceful graph with e edges, then the edges of K_{2e+1} can be partitioned into 2n + 1 graphs isomorphic to G. □

Note that the converse result is not true; for example, there exists a cyclical decomposition of K_{11} into C_5's, although C_5 is not graceful. There exists a wide literature on decompositions of complete graphs (see, for example, [4]). In fact it is sufficient (although again not necessary) to suppose in Theorem 3.1 that G admits a less restrictive numbering (called a ρ-valuation by Rosa [34]). The above conjectures for trees have been popularised by R. A. Duke [11], and a great number of results and references concerning these conjectures and more general graceful graphs appear in the American Mathematical Monthly Research Problems section of R. K. Guy (see [17], [18], [19] and [20]).

It is also worth mentioning that one can consider graceful numberings with additional properties. More precisely, Rosa [34] says that a graph admits an α-valuation if the graph is graceful and if there exists an integer r such that if φ is a graceful numbering for every edge vw of G, then we have either $\phi(v) \le x < \phi(w)$ on $\phi(w) \le x < \phi(v)$. Such a graceful graph is called 'balanced' in [36]. Rosa [34] has proved that not all trees admit an α-evaluation, and Kotzig [27] has conjectured that only a finite number of trees homeomorphic to any given tree do not admit an α-valuation. (The truth of this conjecture would show that almost all trees are graceful and balanced.) Kotzig also proved the following interesting result:

Theorem 3.2. Let T be a given tree. Then the infinite family of trees obtained from T by replacing a given edge by a path of arbitrary length contains only a finite number of trees with no α-valuation. □

We now describe some classes of trees which have been proved to be graceful. The first result is due to Rosa [34]:

Theorem 3.3. All paths are graceful. □

Let us now call a <u>caterpillar</u> a tree such that if we delete the end vertices (vertices of degree 1) then the graph obtained is a path. (In [34] a caterpillar is called a snake.) Rosa proved the following result:

Theorem 3.4. All caterpillars are graceful. □
(Figure 6 shows a caterpillar with a graceful numbering.)

We now use the name <u>lobster</u> for a tree such that if we delete the end-vertices, the graph obtained is a caterpillar (see Figure 7). The first step towards a proof of the first conjecture for trees might be the following natural conjecture:

Conjecture. All lobsters are graceful.

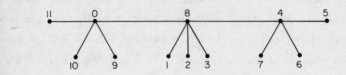

Figure 6

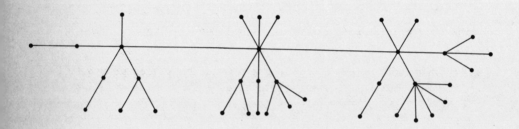

Figure 7

Constructive methods generating graceful trees from smaller ones have been given by many authors; for example, R. G. Stanton and C. R. Zarnke [38], H. Gabow [13], A. Gyárfás and J. Lehel [14], and K. M. Koh, D. G. Rogers and T. Tan [24], [25], [26]. We do not enter in the technical details here; the reader can refer to the survey of Bloom [5], or to any of the articles referenced above. By these methods further classes of trees have been proved to be graceful.

Let us call a tree <u>symmetrical</u> if it is a rooted tree in which every level contains vertices of the same degree. It is proved (for example, in [14]) that:

<u>Theorem 3.5.</u> Symmetrical trees are graceful. ☐

Let us call a <u>complete k-ary tree</u> a special symmetrical tree in which
the degree of the root is k, the degrees of the vertices on other levels
are all k + 1, except those of the vertices of the last level, which are all
1. (In Figure 8, we show a binary tree which is gracefully labelled.)
As a Corollary of Theorem 3.5, we obtain the following result, first proved
in [38]:

<u>Corollary 3.6</u>. Complete k-ary trees are graceful. □

Note that this result has also been rediscovered by many authors, partly
because in 1976 Cahit [10], unaware of Stanton and Zernke's result,
conjectured that binary trees are graceful. Other methods, using the
adjacency matrix of the tree, have been developed by Haggard and McWha [21],
and by Bloom, Haggard and Taylor [8].

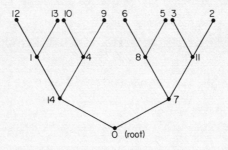

Figure 8

The methods developed can be used to give a graceful labelling to trees
having a minimum 'symmetry'. A. M. Pastel and H. Raynaud [31], [32] have
conversely shown that some 'asymetrical' trees are graceful. Let us call
an <u>olive-tree</u> a rooted tree consisting of k branches, where the ith branch
is a path of length i. It has been proved in [31] that:

Theorem 3.7. All olive-trees are graceful. □

This theorem answers a conjecture of Gyárfás and Lehel (given in a lecture at the Orsay C.N.R.S. Colloquium in 1976). To prove this, Pastel and Raynaud proved that there exists a graceful numbering of the path of length n, such that any value k ($0 \leq k \leq n$) can be taken as the label of one of the end-vertices of the path. The case k = 0 of this last result was first proved by A. Rosa [35]. (Theorem 3.7 has also been proved in a different manner by Koh, Rogers and Tan [24].) In [32] Pastel and Raynaud also exhibited many other families of graceful trees.

To finish the case of graceful trees we mention the following problem due to Kotzig (presented at the Ninth Southeastern Conference on Combinatorics, Boca Raton 1978). To the set of gracefully numbered trees, Kotzig associated the following graph, called G.A.G. in [32]. The vertices of this graph are the graceful trees, gracefully numbered with e edges (two isomorphic trees having different graceful numberings are considered as different). Two vertices are then joined by an edge if the corresponding trees are obtained from each other by deleting one edge and adding another edge with the same label. (See Figure 9 for the case e = 3.) Kotzig conjectured that this graph is 1-factorisable (that is, that it has a perfect matching): this conjecture has recently been proved by Pastel (private communication). Kotzig has conjectured also that the graph G.A.G. is connected.

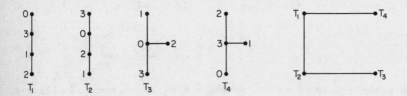

Figure 9

4. COMPLETE GRAPHS

In Figure 1 it was shown that K_3 and K_4 are graceful and it is trivial to see that K_2 can be gracefully labelled. The following result is well known; for a short proof see G. J. Simmons [37, p.630] (see also [7]):

<u>Theorem 4.1.</u> K_n is graceful if and only if $n \leq 4$. □

In view of Theorem 4.1, it is natural to search for the maximum number of edges $e(n)$ of a graceful graph on n vertices. This question was posed by S. W. Golomb [15], [16], but was in fact considered a long time earlier, in terms of the restricted difference basis we met in the Introduction. More precisely, in these terms one wants to know, for a given e, the minimum number of vertices $n(e)$ of a graceful graph with e edges. In the following table we summarize the known results (see the survey of Miller [30] for examples which attain these bounds).

n	3	4	5	6	7	8	9	10	11	12	13	14	15	16	17	18	19	20
e(n)	3	6	9	13	17	23	29	36	43	50	59	60	79	90	101	112	123	138

It has also been proved (see [30], for references) that $\lim_{n \to \infty} e(n)/n^2$ exists, and that $\frac{1}{3} < \lim_{n \to \infty} e(n)/n^2 < 0.411$.

A related problem is that of finding a set of integers $a_1 < a_2 < \ldots < a_n$ such that every positive integer m, $1 \leq m \leq e$, can be represented in the form $m = a_j - a_i$. Such a set is called an <u>unrestricted difference basis</u> with respect to e. The difference between this concept and a restricted difference basis is that here $a_n - a_1$ is not necessarily equal to e. For example, the set $\{0, 8, 18, 19, 22, 24, 31, 39\}$ is an unrestricted difference basis for e = 24; indeed the differences $\{a_j - a_i, j > i\}$ are all the integers from 1 to 24, together with 31 and 39 (8 and 31 appearing twice). Note that by the above table, there does not exist a restricted difference basis with e = 24 and with only 8 terms. The 'unrestricted difference basis problem' consists of finding, for a given e, the minimum value of the number of terms of an unrestricted difference basis with respect to e. This problem is related to a different kind of labelling of a graph.

Let us call a numbering of a graph $\hat{b}$-graceful, if it satisfies conditions (a) and (c) of Section 1. A graph which admits a $\hat{b}$-graceful numbering will be called a $\hat{b}$-graceful graph. The unrestricted difference basis problem is thus equivalent to that of finding, for a given e, the minimum number of vertices of a $\hat{b}$-graceful graph with e edges.

The problem is also equivalent to that of finding for a given n, the maximum number e'_n of edges of a $\hat{b}$-graceful graph of order n. The following table gives the values known. Here also it has been proved that

$$\lim_{n \to \infty} e'(n)/n^2$$

exists, and that

$$\tfrac{3}{8} < \lim_{n \to \infty} e'(n)/n^2 < 0.411$$

(see [30] for details).

n	3	4	5	6	7	8	9	10	11	12	13	14	15	16	17	18	19
e'(n)	3	6	9	13	18	24	29	37	45	51	61	70	79	93	101	113	127

The problem of determining e'(n) may be more difficult than that of determining e(n), although in some cases it is easier to characterise some classes of $\hat{b}$-graceful graphs. For example, it is easy to show by induction on the number of edges that all trees are $\hat{b}$-graceful.

Another related question, motivated by the fact that K_n is not graceful for $n \geq 5$, is the following problem (see S. W. Golomb [15]). One wants to label the vertices of a graph with distinct non-negative integers in such a way that all of the edges are distinctly labelled. (This numbering satisfies property (a) of Section 1, and a property weaker than (c), since we do not require that all integers between 1 and e appear as labels of edges, but only that they are distinct integers.) So the problem is to find the minimum value f(G) of $\phi(v)$ over all vertices of the graph G with such a numbering. For example, for K_5 we know that this minimum is at least 11 (since K_5 is not graceful); in fact $f(K_5) = 11$, since it suffices to label the vertices respectively 0, 1, 4, 9, 11. All ten edges have different labels - namely, the integers from 1 to 11 except 6. The determination of $f(K_n)$ seems to be a very difficult problem, and the values are known only for $n \leq 11$; the known results are given below. For more details and

references, see Simmons [37] (the problem being exactly that of finding
$S(n,1)$ 'synch sets').

n	3	4	5	6	7	8	9	10	11
$f(K_n)$	3	6	11	17	25	34	44	55	72

In view of this table and the known values of $e(n)$, it seems natural to ask whether we always have $e(n) + f(K_n) \geq n(n - 1)$.

We conclude this section by mentioning the following extension of Theorem 4.1, due to Kotzig and Turgeon [28]:

<u>Theorem 4.2</u>. The graph consisting of m vertex-disjoint copies of K_n is graceful if and only if $m = 1$ and $n \leq 4$. □

5. ANTENNAE AND WINDMILLS

To conclude, we mention here how a problem arising in radio-astronomy is related to graceful numbers; for more details see [1] and [2]. The problem is as follows: A few moveable antennae are used in several successive configurations to measure various spatial frequencies relative to some area of sky. Can one obtain an optimal array, where all the distances between antennae are successive multiples of a given increment a without any gaps (otherwise a frequency is missing) and without repetitions (useless)? In practice, a is of the order of the radius of the antennae, and a and 2a cannot themselves be obtained (due to overlapping phenomena).

The combinatorial version can be stated as follows: Let m, n and d be positive integers. Is it possible to find m sets A_i each consisting of n integers a_i^j, $1 \leq j \leq n$, such that if

$$D_i = \{|a_i^j - a_i^{j'}| : 1 \leq j < j' \leq n\},$$

then

$$\sum_{i=1}^{m} D_i = \{k : d \leq k \leq \tfrac{1}{2}mn(n - 1) + d - 1\}.$$

The parameter n corresponds to the number of antennae, m corresponds to the number of days of observation, and d is 1, 2 or 3 (due to the fact that the numbers a and 2a are not obtained).

Example 1.   If $n = 4$, $m = 1$, $d = 1$, we have $A_1 = \{0, 1, 4, 6\}$.

Example 2.   If $n = 4$, $m = 4$, $d = 1$, we have $A_1 = \{0, 1, 7, 23\}$,
$A_2 = \{0, 2, 14, 19\}$, $A_3 = \{0, 3, 13, 21\}$, $A_4 = \{0, 4, 15, 24\}$,
$D_1 = \{1, 6, 7, 16, 22, 23\}$, $D_2 = \{2, 5, 12, 14, 17, 19\}$
$D_3 = \{3, 8, 10, 13, 18, 21\}$, $D_4 = \{4, 9, 11, 15, 20, 24\}$.

Example 3.   If $n = 3$, $m = 3$, $d = 2$, we have $A_1 = \{0, 5, 9\}$, $A_2 = \{0, 6, 8\}$,
$A_3 = \{0, 7, 10\}$.

If $d = 1$, the problem is exactly to determine whether the graph consisting of m K_n's with exactly one vertex in common is graceful. In Figure 10, we show the graceful numbering of 4 K_4's with one vertex in common (as given in Example 2 above).

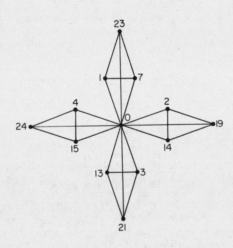

Figure 10

The case $d = 1$, $n = 3$ was a problem raised by C. Hoede [23], and we shall therefore call a <u>Dutch m-windmill</u> the graph consisting of m K_3's with one vertex in common. In fact, in [1] and [2] we solved the problem just before Hoede raised his question, for the solution follows immediately from a solution to Skolem's problem: for which values of m is it possible to partition the integers $\{1, 2, \ldots, 2m\}$ into m pairs $\{a_i, b_i\}$, $i = 1, 2, \ldots, m$, such that $b_i - a_i = i$. The result is:

Theorem 5.1. The Dutch m-windmill is graceful if and only if $m \equiv 0$ or 1 (mod 4). □

For the general case of the problem stated in the Introduction, we proved in [2] the following result which contains Theorem 4.1 as a special case $(m = d = 1)$:

Theorem 5.2. If there exists a solution to the problem, then
either $n = 5$, m is even and $m \geq 4d - 2$;
or $n = 4$ and $m \geq 2d - 1$;
or $n = 3$, $m \geq 2d - 1$ and $m \equiv 0$ or 1 (mod 4) if d is odd,
        $m \equiv 0$ or 3 (mod 4) if d is even. □

If $m = d = 1$, the cases $n = 3$ and $n = 4$ are known (the graceful numberings of K_3 and K_4). If $n = 3$ and $d = 1$, the condition is also sufficient by Theorem 5.1. In [2] we solved many cases for $n = 3$; in particular, the cases $d = 2,3,4$ and m large enough.

For $n = 4$, only a few results are known. For $d = 1$, as we said above, the problem reduces to that of finding a graceful numbering of the graph consisting of m K_4's with exactly one vertex in common; during the Keszthély Colloquium of 1976, it was proposed to call these graphs French m-windmills. The following conjecture appears in [3]:

Conjecture. French m-windmills are graceful if $m \geq 4$.

Figure 10 shows a graceful French 4-windmill. The French 1-windmill is K_4, and is thus graceful. The French 2-windmill is not graceful (see the last graph of Figure 2); M. Gardner [12] has called it a 'particularly ungraceful graph'. The French 3-windmill is also not graceful (see [2] for a proof when $m = 2$ and 3).

Other results concerning the case $d \geq 2$ and $n = 5$ are also given in [2]. Note that when the associated graphs are not graceful, one can (exactly as in Section 4) ask for the maximum number of edges of a graceful subgraph of the graph consisting of m K_n's with exactly one vertex in common.

Let us call a numbering of a graph â-graceful if this numbering is defined as in Section 1, except that it does not necessarily satisfy condition (a). A graph which admits an â-graceful numbering will be called an â-graceful graph. The problem above (in the case d = 1) can be considered as the existence of an â-graceful numbering of the graph consisting of m vertex-disjoint copies of K_n. The examples above (for example, four copies of K_4) show that there exist â -graceful graphs which are not graceful (compare this with Theorem 4.2). If we consider K_8 minus 4 edges, then it can be shown that it is b̂-graceful, but not graceful. It also cannot be â-graceful, because if two vertices of K_8 have the same label, then we can only obtain 21 values for the edges.

There certainly exist â-graceful graphs which are not b̂-graceful, but I am not aware of such an example. As remarked by J. Lehel (private communication), all trees are â-graceful; he has also noted that every â-graceful graph of diameter 2 is graceful, and that Lemma 2.1 is also valid for â-graceful and b̂-graceful graphs.

A related problem, which may be of interest in radio-astronomy (if one wants to move only a restricted number of antennae) is the following:

Problem: For which values of m, n and p is the graph consisting of m K_n's with exactly K_p in common graceful?

The case p = 1 is the above problem (for d = 1). The case n = 4, p = 3 was completely solved by H. Meyniel (private communication), who showed that the graph consisting of m K_4's with a K_3 in common is graceful for every m (see Figure 11 for the corresponding graceful numbering).

Acknowledgement: I wish to thank all the persons who helped me during the preparation of this article - in particular, R. K. Guy, J. Lehel, H. Raynaud, D. G. Rogers, A. Rosa, G. S. Bloom and C. Hoede.

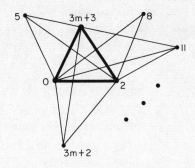

Figure 11

REFERENCES

1. J. C. Bermond, A. E. Brouwer and A. Germa, Systèmes de triples et différences associées, in Colloque C.N.R.S, Problèmes Combinatoires et Théorie des Graphes, Orsay, 1976, C.N.R.S. (1978), 35-38.

2. J. C. Bermond, A. Kotzig and J. Turgeon, On a combinatorial problem of antennas in radio-astronomy, in Colloq. Math. Soc. János Bolyai 18, Combinatorics, Keszthély, Hungary, 1976, North-Holland, Amsterdam (1978), 135-149.

3. J. C. Bermond, Problem in Colloq. Math. Soc. János Bolyai 18, Combinatorics, Keszthély, Hungary, 1976, North-Holland, Amsterdam (1978), 1189.

4. J. C. Bermond and D. Sotteau, Graph decompositions and G-designs, Proc. Fifth British Combinatorial Conference, Aberdeen 1975, Utilitas Math., Congressus Numerantium XV, 53-72.

5. G. S. Bloom, A chronology of the Ringel-Kotzig conjecture and the continuing quest to call all trees graceful, Topics in Graph Theory (ed. F. Harary) (to appear).

6. G. S. Bloom and S. W. Golomb, Applications of numbered undirected graphs, Proc. IEEE 65 (1977), 562-570.

7. G. S. Bloom and S. W. Golomb, Numbered complete graphs, Unusual rulers and assorted applications, in Theory and Applications of Graphs (ed. Y. Alavi and D. R. Lick), Springer-Verlag Lecture Notes in Mathematics 642 (1978), 53-65.

8. G. S. Bloom, G. Haggard and H. Taylor, The adjacency matrices of graceful trees (to appear).

9. H. Bodendiek, H. Schumacher and H. Wegner, Über graziöse Numerierungen von Graphen, Elemente der Mathematik 32 (1977), 49-80.

10. I. Cahit, Are all complete binary trees graceful?, Amer. Math. Monthly 83 (1976), 35-37.

11. R. A. Duke, Can the complete graph with 2n + 1 vertices be packed with copies of an arbitrary tree having n edges? Amer. Math. Monthly 76 (1969), 1128-1130.

12. M. Gardner, Mathematical games, Scientific American 226 (March 1972), 108-112.

13. H. Gabow, How to gracefully number certain symmetric trees, SIGACT News 7 (1975), No.4.

14. A. Gyárfás and J. Lehel, A method to generate graceful trees, in Colloque C.N.R.S. Problèmes Combinatoires et Théorie des Graphes, Orsay, 1976, C.N.R.S. (1978), 207-209.

15. S. W. Golomb, How to number a graph, in Graph Theory and Computing (ed. R. C. Read), Academic Press, New York (1972), 23-37; MR 49-4863.

16. S. W. Golomb, The largest graceful subgraph of the complete graph, Amer. Math. Monthly 81 (1974), 499-501.

17. R. K. Guy, Monthly Research Problems 1969-1973, Amer. Math. Monthly 80 (1973), 1120-1128.

18. R. K. Guy, Monthly Reseach Problems, 1969-1975, Amer. Math. Monthly 82 (1975), 995-1004.

19. R. K. Guy, Monthly Research Problems, 1969-1977, Amer. Math. Monthly 84 (1977), 807-815.

20. R. K. Guy and V. Klee, Monthly Research Problems, 1969-1971, Amer. Math. Monthly 78 (1971), 1113-1122.

21. G. Haggard and P. McWha, Decomposition of complete graphs into trees, Czech. Math. J. 25 (100) (1975), 31-36.

22. C. Hoede and H. Kuiper, All wheels are graceful, Utilitas Math. 14 (1978), 311.

23. C. Hoede Problem in Colloq. Math. Soc. János Bolyai 18, Combinatorics, Keszthély, Hungary 1976, North-Holland, Amsterdam (1978), 1205-1206.

24. K. H. Koh, D. G. Rogers and T. Tan, Two theorems on graceful trees, Discrete Math. (to appear).

25. K. H. Koh, D. G. Rogers and T. Tan, Interlaced trees: a class of graceful trees, 6th Austral. Combinatorics Conf., Armidale, Australia, 1978.

26. K. H. Koh, D. G. Rogers and T. Tan, Another class of graceful trees (to appear).

27. A. Kotzig, On certain vertex-valuations of finite graphs, Utilitas Math. 4 (1973), 261-290.

28. A. Kotzig and J. Turgeon, β-valuations of regular graphs with complete components, in Colloq. Math. Soc. János Bolyai 18, Combinatorics, Keszthély, Hungary, 1976, North-Holland, Amsterdam (1978), 697-703.

29. J. Leech, On the representation of 1, 2, ..., n by differences, J. London Math. Soc. 31 (1956), 160-169; MR 19-942.

30. J. C. P. Miller, Difference bases; three problems in additive number theory, in Computers and Number Theory (ed. A. O. L. Atkin and B. J. Birch), Academic Press (1971), 299-322; MR 47-4817.

31. A. M. Pastel and H. Raynaud, Les oliviers sont gracieux, in Colloq. Grenoble, 1978, Publications Université de Grenoble.

32. A. M. Pastel and H. Raynaud, Rapport de recherche, LA7 IMAG.

33. G. Ringel, Problem 25 in Theory of graphs and its applications, Proc. Symposium Smolenice 1963, Prague (1964), 162.

34. A. Rosa, On certain valuations of the vertices of a graph, in Théorie des Graphes, Rome 1966 (ed. P. Rosenstiehl), Dunod, Paris (1968), 349-355; MR 36-6319.

35. A. Rosa, Labeling snakes, Ars Combinatoria 3 (1977), 67-74.

36. D. A. Sheppard, The factorial representation of balanced labelled graphs, Discrete Math. 15 (1976), 379-388.

37. G. J. Simmons, Synch sets: A variant of difference sets, Proceeding of the Fifth Southeastern Conf. on Combinatorics, Graph Theory and Computing, Boca Raton 1974, Utilitas Math., Congressus Numerantium X (1974), 625-645; MR 50-9628.

38. R. G. Stanton and C. R. Zarnke, Labelling of balanced trees, Proc. Fourth Southeastern Conf. on Combinatorics, Graph Theory and Computing, Boca Raton 1973, Utilitas Math. Congressus Numerantium VIII (1973), 479-495; MR 51-2959.

J.-C. Bermond
Informatique, bât 490
Université Paris Sud
91405 - Orsay
France

P J Cameron
Multiple transitivity in graphs

The concept of multiple transitivity has been of fundamental importance in the theory of permutation groups for a century. In this article, I define "multiple transitivity" for a variety of mathematical structures. (Essentially, a structure is 'k-transitive' if each isomorphism between substructures on at most k points is induced by an automorphism.) When applied to graphs, a surprising result emerges: depending on the precise definitions used, all 5-transitive or 6-transitive graphs are trivial!

Let G be a group of permutations on a set X, and let k be a positive integer. We say that G is <u>k-transitive</u> if, given any two k-tuples $(x_1,\ldots,x_k)$ and $(y_1,\ldots,y_k)$ of distinct elements of X, there is an element $g \in G$ for which $x_i g = y_i$ for $i = 1,\ldots,k$. (Here $x_i g$ denotes the image of x_i under the permutation g.) To avoid trivialities, it is usually assumed that $k \leq |X|$; with this condition, we see easily that a k-transitive group is ℓ-transitive for all $\ell \leq k$. The symmetric group on X is n-transitive (where $|X| = n$), and the alternating group is (n-2)-transitive; these may be regarded as 'trivial'. Apart from these groups, there is no known finite 6-transitive group, and only two known 5-transitive groups, the two <u>Mathieu groups</u> M_{12} and M_{24}. It is still an open problem to show that there are no more. (Though the result mentioned in the preceding paragraph looks tantalisingly similar, there appears to be no connection.)

The definition of k-transitivity can be rephrased thus: every bijection between subsets of X of cardinality k is induced by an element of G. This is the basis of our generalization.

Let $\underline{X}$ be a class of mathematical structures, satisfying the following two conditions:

(i) every member X of $\underline{X}$ has an underlying set of "points" (and, following universal mathematical usage, we denote this set of points by X also);

(ii) for any $Y \subseteq X$, there is an 'induced substructure' of X with point set Y. (This substructure need not itself belong to the class $\underline{X}$, although it is more convenient if it does.)

For a given class $\underset{\sim}{X}$, there may be more than one choice of what we mean by "induced substructure"; different choices may give different theories of multiple transitivity, as we shall see in the case of graphs. Suitable classes include

(a) topological spaces (with the usual definition of subspace);
(b) linearly, partially or circularly ordered sets;
(c) matroids (the induced matroid on Y may be taken to be either the restriction or the contraction to Y);
(d) graphs (see later).

For projective spaces, the most natural definition of substructure seems to be the induced submatroid, so this case is best included under (c). For groups, there is no satisfactory concept.

Now our definition of k-transitivity reads as follows. A member X of the class $\underset{\sim}{X}$ is said to satisfy condition A_k if any isomorphism between induced substructures on at most k points extends to an automorphism of X. The words "at most" are included to ensure that a property noted earlier for k-transitivity holds:

Proposition 1. A_k implies A_ℓ whenever $\ell \leq k$. □

We can now pose the question: given a class $\underset{\sim}{X}$ and a positive integer k, which members of $\underset{\sim}{X}$ satisfy A_k? The answer may be trivial. For example:

Proposition 2. No finite partially ordered set can ever satisfy A_1, except for the trivial case when any two elements are incomparable.

Proof. No automorphism can map a minimal element to one which is not minimal. □

On the other hand, the question may be so difficult as to be beyond reach. We will see that graphs provide an intermediate case.

The next (well-known) result illustrates a surprising phenomenon which we will encounter again later. Note that, in contrast to Proposition 1, infinite ordered sets (such as the real numbers) may satisfy A_k.

Proposition 3. An infinite totally ordered set satisfying A_2 also satisfies A_k for every positive integer k.

Proof. Let $\{x_1,\ldots,x_k\}$ and $\{y_1,\ldots,y_k\}$ be two k-sets; we may suppose that $x_1 < \ldots < x_k$ and $y_1 < \ldots < y_k$. Then the unique isomorphism θ between these two sets maps x_i to y_i for $1 \leq i \leq k$. By A_2, there is an isomorphism θ_i from $\{x : x_i < x < x_{i+1}\}$ to $\{x : y_i < x < y_{i+1}\}$ for $1 \leq i \leq k-1$. Also, since A_1 holds, there are isomorphisms

$$\theta_0 : \{x : x < x_1\} \longrightarrow \{x : x < y_1\}$$
$$\text{and } \theta_k : \{x : x > x_k\} \longrightarrow \{x : x > y_k\}.$$

We have now defined piecewise an automorphism θ' of X extending θ. □

Automorphism groups of totally ordered sets have been studied in depth; we refer to [14] for a survey. A feature of the theory is that several concepts of permutation group theory other than multiple transitivity have been extended to this situation. In passing, we note the surprising result of [4], which goes in the opposite direction to Proposition 3.

We now turn to the main subject of this paper, graph theory. All our graphs will be finite and simple. What is meant by "induced substructure"? An obvious candidate is the induced subgraph. This approach has been pioneered by Gardiner [11]. He calls a graph Γ <u>ultrahomogeneous</u> if every isomorphism between induced subgraphs extends to an automorphism of Γ; that is, if A_k holds for all positive integers k. He has determined all ultrahomogeneous graphs. (This result will appear later, as a consequence of Theorem 5.)

However, there is another possibility, motivated by the work of Biggs and his school. Any connected graph Γ gives rise to a metric d on its vertex-set: the distance between two vertices is (as usual) the length of the shortest path joining them. The graph Γ is called <u>distance-transitive</u> if, given any two pairs (x_1,x_2) and (y_1,y_2) of vertices with $d(x_1,x_2) = d(y_1,y_2)$, there is an automorphism of Γ mapping x_1 to y_1 and x_2 to y_2. Distance-transitive graphs are discussed extensively in Biggs [2]. Meredith [18] extended the concept by calling Γ <u>k-tuple transitive</u> if any isometry between k-tuples of vertices extends to an automorphism of Γ. (Unlike Meredith, I allow k-tuples to have repeated entries; this ensures that, as usual, k-tuple transitivity implies ℓ-tuple transitivity for $\ell \leq k$.) 2-tuple transitivity is precisely the same as distance-transitivity.

We now have two concepts of multiple transitivity for graphs. If $\underset{\sim}{X}$ is the class of graphs, and "induced substructure" means "induced subgraph", condition A_k will be denoted by A_k (induced). If $\underset{\sim}{X}$ is the class of connected graphs, and "induced substructure" means "metric subspace", the condition will be called A_k(metric); this is k-tuple transitivity. We record some elementary facts about these concepts.

Proposition 4. (i) For connected graphs, A_k(induced) implies A_k(metric) for all k, and is equivalent to it for k = 1.
(ii) A connected graph satisfying A_2(induced) has diameter at most 2, while a disconnected graph satisfying A_2(induced) is a disjoint union of complete graphs.
(iii) If a graph satisfies A_k(induced), so does its complement.

Proof. (i) Since any isometry between subsets is necessarily an isomorphism of induced subgraphs, there are fewer isometries, and the metric condition puts fewer demands on the graph.
(ii) Suppose Γ satisfies A_2(induced). If there exists a pair of vertices at distance 2, then any two non-adjacent vertices have distance 2, and Γ is connected with diameter 2. If not, then every component of Γ is a complete graph. (Because of this, we may consider only connected graphs.)
(iii) Clear. □

We come now to the main results.

Theorem 5. A graph satisfying A_5(induced) is one of the following:
(i) a disjoint union of complete graphs of the same size;
(ii) a regular complete multipartite graph;
(iii) the pentagon;
(iv) the line graph of $K_{3,3}$. □

Corollary 6. A graph satisfying A_5(induced) also satisfies A_k(induced) for all positive integers k; and 5 is best possible. □

Theorem 7. A connected graph satisfying A_6(metric) is one of the following:
(i)   a complete multipartite graph;
(ii) a complete bipartite graph with the edges of a matching deleted;
(iii) a cycle;
(iv) the line graph of $K_{3,3}$;
(v)   the icosahedron;
(vi) a unique graph of valency 9 on 20 vertices. (The vertices are the 3-subsets of a 6-set, joined if they meet in a 2-set.) □

Corollary 8. A connected graph satisfying A_6(metric) satisfies A_k(metric) for all positive integers k; and 6 is best possible. □

Gardiner's theorem on ultrahomogeneous graphs is a corollary of Theorem 5. We outline the proofs of the theorems, without giving the calculations. Details will appear elsewhere [6].

The ideas in the proof of Theorem 5 can be said to originate in the work of Higman [13] on "rank 3 permutation groups"; this can be regarded as being about graphs which satisfy A_2(induced). (By convention, disjoint unions of complete graphs and complete multipartite graphs are excluded here.) Higman showed that the number of vertices, valency, and other numerical functions of such graphs can be expressed in terms of three parameters, which are independent but satisfy a certain "integrality condition". Margaret Smith [23] used matrix-theoretic methods to study graphs satisfying A_3(induced), and found that for these graphs, Higman's expressions simplify to two 2-parameter families.

In a graph Γ, the subconstituents Γ(x) and Δ(x) with respect to a vertex x are the induced subgraphs on the sets of vertices adjacent and non-adjacent to x, respectively. The key observation is that, if Γ satisfies A_k(induced), then both subconstituents satisfy A_{k-1}(induced). Thus, if Γ satisfies A_4(induced), then both Γ and its subconstituents are described by Smith's formulae; calculation shows that only $L(K_{3,3})$ and a 1-parameter family are possible. Repeating the argument for A_5(induced) shows that, of this 1-parameter family, only the pentagon survives. This proves Theorem 5.

The proof of Theorem 7 is similar. If Γ satisfies A_k(metric), then the subconstituent Γ(x) satisfies A_{k-1}(induced), since two non-adjacent vertices of $\Gamma(x)$ are at distance 2 in Γ. If $\Gamma(x)$ is a null graph, then Γ has girth greater than 3. This case was considered by Meredith [18], who showed that a graph with girth greater than 4 which satisfies A_3(metric) must be a cycle. A slight modification of his argument determines all graphs of girth 4 satisfying A_4(metric). Thus, if Γ satisfies A_6(metric), we may assume that $\Gamma(x)$ is non-null and is one of the graphs of Theorem 5. In each case, the graph Γ can be determined; only type (i) in Theorem 5 gives any difficulty.

The corollaries are proved by examining the graphs in the theorems, together with the following examples to show that the numbers 5 and 6 are best possible. The "Schläfli graph" on 27 vertices, associated with the 27 lines in a general cubic surface, and its complement, both satisfy A_4(induced) but not A_5(induced); see [23]. The graphs obtained from these two by the construction of Shult's "graph extension theorem" [22] both satisfy A_5(metric) but not A_6(metric).

From Theorem 5, we can deduce a similar result for uniform hypergraphs. Let $\underset{\sim}{X}$ be the class of finite simple s-uniform hypergraphs (whose edges are distinct s-subsets of the vertex-set X). The induced substructure on Y has all those edges of X which are contained in Y.

Theorem 9. An s-uniform hypergraph ($s \geq 3$) satisfying A_{s+3} is one of the following:
(i)   the complete or null hypergraph;
(ii) the hypergraph whose edges are the lines of PG(2,2), or its complement;
(iii) the hypergraph whose edges are the planes of AG(3,2), or its complement;
(iv) the unique regular two-graph on 6 points;
(v)   the unique regular two-graph on 10 points.
(Here PG(2,2) and AG(3,2) denote the projective plane and affine 3-space over GF(2). For regular two-graphs, see [21].) □

Corollary 10. An s-uniform hypergraph ($s \geq 2$) satisfying A_{s+3} satisfies A_k for all positive integers k. □

We conclude with a number of remarks and open problems raised by these results.

It is easy to find examples to show that the conditions A_k(induced) and A_k(metric) become strictly stronger up to the bounds given by Corollaries 6 and 8. This suggests the problem of determining or characterizing these classes of graphs, especially A_4(induced) and A_5(metric). (The examples mentioned earlier for these classes are the only ones known.) A related problem: can the bound in Corollary 10 be improved for sufficiently large s?

As is well known, any t-transitive permutation group (with a few exceptions, namely those of [1]) is a group of automorphisms of a non-trivial t-design; the reason is very simple - see [15]. In [9], Delsarte, Goethals and Seidel defined the concept of a "spherical t-design". This is a finite set of points on the Euclidean unit sphere in R^d, whose distribution has a certain uniformity: precisely, for $1 \leq i \leq t$, the i-th moments of the set are invariant under rotations of the sphere. Although this definition looks very different from that of an ordinary t-design, there are a number of subtle and remarkable parallels: see [8], for example. One such case occurs here. From any strongly regular graph, and in particular any graph satisfying A_2(induced), two spherical 2-designs can be constructed in a canonical way: they are the images of the orthonormal basis labelled by the vertices under orthogonal projection onto eigenspaces (other than that spanned by the all-1 vector) of the adjacency matrix of the graph. These have the additional property that only two angular distances between points appear. It turns out that for one of Smith's families of parameters of graphs satisfying A_3(induced), a spherical 3-design is obtained. Furthermore, the 1-parameter family satisfying A_4(induced) yields spherical 4-designs. Since no spherical 5-design can have just two angular distances (see [9]), we expect the procedure to stop here. It seems to be important to understand the precise relationship between multiple transitivity and spherical designs. The paper [7] throws some light on the case k = 3. A closely related phenomenon is mentioned in [5].

Another interesting question is that of finding direct proofs of the corollaries, not depending on inspection of the lists in the theorems. (Bear in mind the simple proof of Proposition 3.) Perhaps a satisfactory answer to the last problem whould help here.

We turn now to some questions of a more general nature. Can other concepts of permutation group theory be usefully extended to our set-up? As we have noted, this has been done successfully for totally ordered sets. There are also some results for graphs, which we now discuss.

A permutation group G on a set X is called <u>k-homogeneous</u> if, given any two k-subsets of X, there is an element of G mapping the first to the second (in some order). The principal result on such finite groups is due to Livingstone and Wagner [17]: if $|X| \geq 2k$, a k-homogeneous group on X is (k-1)-transitive, and is k-transitive if $k \geq 5$. (For infinite groups, the position is different: see [4].) The group G is called <u>generously (k-1)-transitive</u> if, given any k-subset Y of X, each permutation of Y is induced by an element of G. A generously (k-1)-transitive group, finite or infinite, is (k-1)-transitive [18] - hence the name! A group is k-transitive if and only if it is both k-homogenous and generously (k-1)-transitive.

Let us define analogous concepts for graphs. We will deal only with the induced-subgraph formalism, and so the word "<u>(induced)</u>" will be omitted. We say that the graph Γ satisfies <u>weak</u> A_k if, whenever two induced subgraphs on at most k vertices are isomorphic, then some automorphism of Γ maps the first to the second. Also, Γ satisfies <u>local</u> A_k if every automorphism of any induced subgraph on at most k vertices extends to an automorphism of Γ. Now Γ satisfies A_k if and only if it satisfies both weak and local A_k.

It has been shown by Ronse [20] and Gardiner [12] that, if Γ satisfies either weak or local A_k for all k, then it satisfies A_k for all k, and the conclusions of Theorem 5 hold. These results can be improved in the same way that Theorem 5 improves Gardiner's Theorem [11].

<u>Theorem 11</u>. (a) If Γ is a graph satisfying weak A_5, then Γ satisfies A_k for all k (and the conclusions of Theorem 5 hold).
(b) For any graph Γ, local A_k implies A_{k-1}. Hence, if Γ satisfies local A_6, then Γ satisfies A_k for all k (and the conclusions of Theorem 5 hold).

<u>Proof</u> (in outline). (a) The proof of Theorem 5 given in [6] is combinatorial, and does not use the full force of A_5. It can be shown (using the inclusion-exclusion principle) that weak A_5 implies the

combinatorial hypotheses used in the proof.

(b) This is proved by induction on k. If local A_2 holds, then any two vertices can be interchanged by an automorphism, and the automorphism group is transitive. Suppose that local A_k holds, and (by induction) that A_{k-2} holds. Let $v_1,\ldots,v_{k-1}$, $w_1,\ldots,w_{k-1}$ be vertices such that the map $v_i \to w_i$ ($1 \le i \le k-1$) is an isomorphism of induced subgraphs. Using A_{k-2}, we may assume that $v_i = w_i$ for $1 \le i \le k-2$. Now the induced subgraph on $\{v_1,\ldots,v_{k-1},w_{k-1}\}$ has an automorphism fixing $v_1,\ldots,v_{k-2}$ and interchanging v_{k-1} and w_{k-1}. By local A_k, this is induced by an automorphism of Γ. □

Some questions remain. For example, is it true that weak A_k implies A_k for all k? (Only k = 2,3,4 are undecided.)

Another modification considered by Gardiner is to assume only that isomorphism between induced subgraphs of certain specified types extend to automorphisms of Γ. It would be worth trying to combine the ideas of [12] and [6]. Mention should also be made here of Tutte's concept of arc-transitivity [24]. A similar example from a different field is basis-transitivity for matroids.

Finally, there is the question of discovering similar (non-trivial) results for other classes of mathematical structures. As we have seen, the first major difficulty is the choice of structure (and induced sub-structure). A good candidate appears to be the class of directed graphs. (As well as excluding loops and multiple edges, we will prohibit the occurrence of edges joining the same vertices in opposite directions.)

For tournaments, there is a satisfactory answer. A <u>Paley tournament</u> has as vertex-set the finite field $GF(q)$, where $q \equiv 3 \pmod 4$; the edges are those pairs (x,y) for which $x-y$ is a non-zero square. The following theorem is due to Kantor [16]:

<u>Theorem 12</u>. Any tournament satisfying A_2 is a Paley tournament. The only tournament satisfying A_3 is the directed 3-cycle. □

(Kantor's proof uses the Feit-Thompson theorem [10]; the problem of finding an elementary proof is still open.)

Some trivial directed graphs satisfying A_2 can be constructed from tournaments (and A_3 holds if the directed 3-cycle is used):

(i) take the disjoint union of m copies of a Paley tournament;

(ii) replace each vertex x of a Paley tournament by m vertices x_i ($1 \leq i \leq m$), and each edge (x,y) by m^2 edges (x_i, y_j), where $1 \leq i, j \leq m$.

This is also an interesting digraph on 36 vertices with valency 7, satisfying A_2 (see [3]). But I conjecture that no no-null digraphs satisfy A_3 except for those constructed as above from the directed 3-cycle; these incidentally satisfy A_k for all k.

Other classes of structures which could be studied include graphs with loops and multiple edges, edge-coloured graphs, association schemes, and non-uniform hypergraphs (in particular, simplicial complexes).

REFERENCES

1. R. A. Beaumont and R. P. Peterson, Set-transitive permutation groups, Canad. J. Math. 7 (1955), 35-42; MR 16-793.

2. N. L. Biggs, Algebraic Graph Theory, Cambridge University Press, Cambridge, 1974; MR 50-151.

3. P. J. Cameron, Primitive groups with most suborbits doubly transitive, Geometriae Dedicata 1 (1973), 436-446; MR 48-391.

4. P. J. Cameron, Transitivity of permutation groups on unordered sets, Math. Z. 148 (1976), 127-139.

5. P. J. Cameron, Extensions of designs: variations on a theme, Combinatorial Surveys (ed. P. J. Cameron), Academic Press, London (1977), 22-42.

6. P. J. Cameron, 6-transitive graphs, J. Combinatorial Theory (B), to appear.

7. P. J. Cameron, J. -M. Goethals and J. J. Seidel, Strongly regular graphs with strongly regular subconstituents, J. Algebra, to appear.

8. P. Delsarte, Hahn polynomials, discrete harmonics, and t-designs, SIAM J. Appl. Math. 34 (1978), 157-166.

9. P. Delsarte, J. -M. Goethals and J. J. Seidel, Spherical codes and designs, Geometriae Dedicata 6 (1977), 363-388.

10. W. Feit and J. G. Thompson, Solvability of groups of odd order, Pacific J. Math. 13 (1963), 775-1029; MR 29-3538.

11. A. D. Gardiner, Homogeneous graphs, J. Combinatorial Theory (B) 20 (1976), 249-257.

12. A. D. Gardiner, Homogeneity conditions in graphs, J. Combinatorial Theory (B) 24 (1978), 301-310.

13. D. G. Higman, Finite permutation groups of rank 3, Math. Z. 86 (1964), 145-156; MR 32-4182.

14. W. C. Holland, Ordered permutation groups, Permutations (Actes du Colloq., Paris, 1972), Gauthier-Villars, Paris (1974), 57-64; MR 49-8914.

15. D. R. Hughes, On t-designs and groups, Amer. J. Math 87 (1965), 761-778; MR 32-5727.

16. W. M. Kantor, Automorphism groups of designs, Math. Z. 109 (1969), 246-252; MR 43-71.

17. D. Livingstone and A. Wagner, Transitivity of finite permutation groups on unordered sets, Math. Z. 90 (1965), 393-403; MR 32-4183.

18. G. H. J. Meredith, Triple transitive graphs, J. London Math. Soc. (2) 13 (1976), 249-257.

19. P. M. Neumann, Generosity and characters of multiply transitive permutation groups, Proc. London Math. Soc. (3) 31 (1975), 457-481; MR 52-3291.

20. C. Ronse, On homogeneous graphs, J. London Math. Soc. (2) 17 (1978), 375-379.

21. J. J. Seidel, A survey of two-graphs, Teorie Combinatorie, Tome I, Accad. Naz. Lincei, Roma (1976), 481-511.

22. E. E. Shult, The graph extension theorem, Proc. Amer. Math. Soc. 33 (1972), 278-284; MR 45-3547.

23. M. S. Smith, On rank 3 permutation groups, J. Algebra 33 (1975), 22-42; MR 52-5775.

24. W. T. Tutte, On the symmetry of cubic graphs, Canad. J. Math. 11 (1959), 621-624; MR 22-679.

Peter J. Cameron
Merton College
Oxford OX1 4JD
England

M Gordon and J A Torkington
Improvements in the random walk model of polymer chains

1. INTRODUCTION

The model (Figure 1) of a polymer chain consisting of a string of n beads with pairs of beads mutually touching to form contacts is frequently used. The counting of configurations available to the chain can be facilitated by the correspondence between chain configurations and lattice walks. One end-bead of the chain is placed at a point of, say, the diamond lattice graph, and successive beads then mark other points on the lattice visited by the 'walker'.

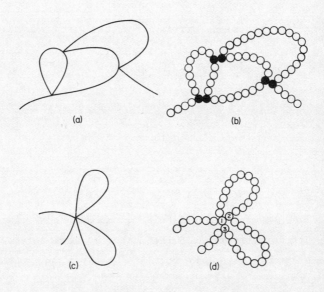

Figure 1

Every contact pair has an associated interaction energy E, and the canonical Gibbs form of the partition function of the polymer chain arises naturally as a function of the number n_m of configurations (lattice walks) which have exactly m contact pairs. The partition function for a chain of n beads with $\beta = E/kT$ is written

$$Z(n) = \sum_{m=0}^{P(n)} n_m(n) e^{-m\beta}, \tag{1}$$

where $P(n)$ is the maximum number of contact pairs possible, and the $\exp(-m\beta)$ is the Boltzmann factor.

If r_m denotes a physical property over all chains with precisely m contact pairs, then that property is calculated

$$r = \sum_{m=0}^{P(n)} n_m r_m e^{-m\beta} / Z(n). \tag{2}$$

Another form of the partition function has been widely used (see, for example, Fixman [4], and Domb and Joyce [2]), but its use in polymer science has been beset by divergence troubles, although in other fields it is often of great value (see Preston [8]). We now derive the discrete form of this alternative partition function. Write

$$Z(n) = \sum_{m=0}^{P(n)} n_m(n) [1 - (1 - e^{-\beta})^m], \tag{3}$$

and expand in powers of $1 - \exp(-\beta)$. Then

$$Z(n) = \sum_{m=0}^{P(n)} [-(1 - e^{-\beta})]^m [\binom{m}{m} n_m + \binom{m+1}{m} n_{m+1} + \ldots + \binom{P(n)}{m} n_{P(n)}]. \tag{4}$$

Recent mathematical theory (see Flatto [5], and Jain and Pruitt [7]) shows that the asymptotic distribution of the number of self-intersections in a random walk on a lattice takes the form

$$n_m(n) \sim (\text{const.}) e^{-[(m - \mu_0(n))^2 / 2\sigma_0^2(n)]}, \text{ as } n \to \infty. \tag{5}$$

Here a return of the walker to a point already visited is called a <u>self-intersection</u>, and $\mu_0(n)$ is the mean number of self-intersections. It is given by

$$\mu_0(n) = \rho n - O(n^{\frac{1}{2}}), \tag{6}$$

while remarkably the variance

$$\sigma_0^2(n) \sim n \log n, \text{ as } n \to \infty. \tag{7}$$

The constant ρ is the <u>inner probability</u> of the lattice walk. We propose to interpret a self-intersection of a walk as contact of a pair of beads in the corresponding polymer chain.

2. THE MEAN SQUARE RADIUS OF A POLYMER CHAIN

The physical property of greatest interest is probably the mean square radius, say $\overline{r^2}$, which is the mean square distance of any point on the chain from the centre of mass. Since n_m is known to be asymptotically Gaussian (by (5)), and β is the parameter to be adjusted, all is available on the right of (2) except how $\overline{r_m^2}$ varies with m. Light scattering is an experimental method of measuring $\overline{r^2}$, and for a random walk it is known that $\overline{r^2} = \frac{1}{6}nb^2$, where b is the step length, regardless of the embedding space. However, the problem of calculating $\overline{r_m^2}$ seems intractable. Not surprisingly, the estimation of the mean square radius of all chains which have at least a given set of m contact pairs is an easier proposition. Let $\overline{R_m^2}$ be the mean square radius of all chains which have at least a given set of m contact pairs averaged over all possible sets of m contact pairs. It is easy to see that

$$\overline{R_m^2} = \frac{n_m \overline{r_m^2} + n_{m+1} \overline{r_{m+1}^2} \binom{m+1}{1} + n_{m+2} \overline{r_{m+2}^2} \binom{m+2}{2} + \cdots}{n_m + n_{m+1} \binom{m+1}{1} + n_{m+2} \binom{m+2}{2} + \cdots}. \tag{8}$$

Define the diagonal matrices

$$[\overline{R^2}] = \text{diag}\{\overline{R_0^2}, \overline{R_1^2}, \overline{R_2^2}, \ldots, \overline{R_{P(n)}^2}\}, \tag{9}$$

$$[\overline{r^2}] = \text{diag}\{\overline{r_0^2}, \overline{r_1^2}, \overline{r_2^2}, \ldots, \overline{r_{P(n)}^2}\}, \tag{10}$$

$$\underline{n}^T = [n_0, n_1, n_2, \ldots, n_{P(n)}], \tag{11}$$

where T means transposition, and the matrix

$$[P] = \begin{bmatrix} \binom{0}{0} & \binom{1}{0} & \binom{2}{0} & \binom{3}{0} & \cdots \\ 0 & \binom{1}{1} & \binom{2}{1} & \binom{3}{1} & \cdots \\ 0 & 0 & \binom{2}{2} & \binom{3}{2} & \cdots \\ 0 & 0 & 0 & \binom{3}{3} & \cdots \\ \cdot & \cdot & \cdot & \cdot & \cdots \end{bmatrix} \equiv \begin{bmatrix} 1 & 1 & 1 & 1 & \cdots \\ 0 & 1 & 2 & 3 & \cdots \\ 0 & 0 & 1 & 2 & \cdots \\ 0 & 0 & 0 & 1 & \cdots \\ \cdot & \cdot & \cdot & \cdot & \cdots \end{bmatrix} . \quad (12)$$

P is, of course, closely related to Pascal's triangle, and has the following matrix inversion property (see Comtet [1, p.143]):

$$[P]^{-1} = \begin{bmatrix} +\binom{0}{0} & -\binom{1}{0} & +\binom{2}{0} & -\binom{3}{0} & \cdots \\ 0 & +\binom{1}{1} & -\binom{2}{1} & +\binom{3}{1} & \cdots \\ 0 & 0 & +\binom{2}{2} & -\binom{3}{2} & \cdots \\ 0 & 0 & 0 & +\binom{3}{3} & \cdots \\ \cdot & \cdot & \cdot & \cdot & \cdots \end{bmatrix} . \quad (13)$$

Now we can write

$$[\overline{R^2}][P]\underline{n} = [P][\overline{r^2}]\underline{n}, \quad (14)$$

and therefore

$$[P]^{-1}[\overline{R^2}][P]\underline{n} = \overline{r^2}\, \underline{n},$$

or in component form

$$\overline{r_m^2} = \frac{1}{n_m} \sum_{i=m}^{P(n)} \binom{i}{i-m}(-1)^{m+i}\, \overline{R_i^2} \sum_{j=i}^{P(n)} \binom{j}{j-i} n_j. \quad (15)$$

Asymptotically as $n \to \infty$, the behaviour of this violently oscillating series is undefined, just like (3), but in applications to physical measurements, this behaviour is damped out by the Gaussian weights, and when β is positive in addition by Boltzmann weights. Defining the vector

$$\underline{B}^T = [e^{-0\beta}, e^{-1\beta}, e^{-2\beta}, \ldots, e^{-P(n)\beta}], \tag{16}$$

we may now write

$$\begin{aligned}\overline{r^2} &= \underline{B}^T[\overline{r^2}]\underline{n} / \underline{B}^T\underline{n} \\ &= \underline{B}^T[P]^{-1}[\overline{R^2}][P]\underline{n} / \underline{B}^T\underline{n}.\end{aligned} \tag{17}$$

The expansion of $\underline{B}^T[\overline{r^2}]\underline{n} / \underline{B}^T\underline{n}$ has no negative terms and readily converges. However, the evaluation of $[\overline{r^2}] = \text{diag}\{\overline{r_0^2}, \overline{r_1^2}, \ldots\}$ is so difficult that hundreds of papers have been written on the first diagonal term $\overline{r_0^2}$ alone. The right-hand expression has $[\overline{R^2}]$ instead of $[\overline{r^2}]$, which is rather easier, but requires the inversion procedure using Pascal's triangle. (In his Pensée 348, Pascal himself said: through space, the Universe grasps and engulfs me like a point; through thought, I grasp it.)

3. ESTIMATION OF $\overline{R_m^2}$ (FIRST APPROXIMATION)

For unconstrained random lattice walks, the mean-square radius is asymptotically one-sixth of the mean-square end-to-end distance, which in general is easier to calculate. This proportionality is discussed in Yamakawa [9]. We use it hereafter as an approximation for constrained graphs.

Figure 2

Consider the graph of a lattice walk known to have at least a given m contacts (Figure 2(a)). Submit to elementary contraction all lines (bonds) except those in the two terminal sections of the chain (Figure 2(b)). The contracted graph represents an unconstrained random walk whose mean-square length is asymptotically equal to νb^2, where ν is the number of bonds and b is the bond length. It remains to find the average over ν for all choices of the 2m beads out of n (the number of beads or chain length) for making the m contacts. If these beads are chosen at random along the chain, then the mean number of bonds of the contracted chain, comprising the terminal sections of the original uncontracted chain, is given thus:

$$\bar{\nu} = \frac{n}{m + \frac{1}{2}}, \text{ for } m = 1, 2, \ldots, \tag{18}$$

and $\bar{\nu} = \frac{n}{m + 1}$, trivially for $m = 0$. (19)

Since the mean value μ_0 of the asymptotically Gaussian n_m distribution is large, we may smooth (18) and (19) by taking (19) to be valid for all m, so that

$$\overline{R_m^2} = \tfrac{1}{6}b^2\,\bar{\nu} = \frac{b^2 n}{6(m + 1)}. \tag{20}$$

We have good reason to guess that $\overline{R_m^2} > b^2 n/6(m + 1)$, since the elementary contraction process is bound to reduce the value of $\overline{R_m^2}$. In greater physical detail, the 2m beads for pairing into m contacts should not have been picked at random along the chain graph, but each set chosen should be weighted by the number of distinct lattice walks in which it occurs. Since intuitively these weights favour more extended chains, the conclusion that for large n

$$\overline{R_m^2} \geq \frac{b^2 n}{6(m + 1)} \tag{21}$$

is a strong conjecture. Note that the solution is graph-theoretical, and leans on the dimensionality of the embedding space only through the first equality of (20) which is true for random walks embedded in spaces of any number of dimensions.

Edwards and Freed [3] were concerned with the mean square radii of homodisperse chains 'cyclised' – that is, constrained by inserting at random m permanent bonds (permanent contacts)–into a chain of n repeat units. Their method, being based on the self-consistent field perturbation method and within a three-dimensional continuum framework, is the one that will most commend itself to physicists, and gives

$$\overline{R^2_m} \sim \frac{nb^2}{6(m+1)}$$

Four other methods are given in Gordon et al.[6].

4. THE INVERSION

The Möbius-type inversion formula of (16), may be expanded in the form

$$\overline{r^2} = \sum_{m=0}^{P(n)} n_m(n) \sum_{t=0}^{m} \sum_{p=0}^{t} e^{-p\beta} \binom{m}{t}\binom{t}{p}(-1)^{t-p} R_t^2 / Z(n). \quad (22)$$

Substituting our crude approximation to $\overline{R^2_m}$ gives

$$\alpha^2 = \frac{6\overline{r^2}}{nb^2} = \sum_{m=0}^{P(n)} n_m(n) \sum_{t=0}^{m} \sum_{p=0}^{t} \frac{1}{t+1} e^{-p\beta} \binom{m}{t}\binom{t}{p}(-1)^{t-p}/Z(n), \quad (23)$$

where α^2 is known as the 'expansion factor'.

The summation over t and p are found combinatorially by the trinomial expansion:

$$(e^{-\beta} - 1 + x^{-1})^m \equiv \sum_{t=0}^{m} \sum_{p=0}^{t} e^{-p\beta} \binom{m}{t}\binom{t}{p}(-1)^{t-p} x^{-m+t}. \quad (24)$$

If we multiply both sides by x^m, collect factorials into binomial coefficients, and integrate over x from 0 to 1, we get

$$\sum_{t=0}^{m} \sum_{p=0}^{t} \frac{1}{t+1} e^{-p\beta} \binom{m}{t}\binom{t}{p}(-1)^{t-p} = \int_0^1 (1 - x(1 - e^{-\beta}))^m dx \quad (25)$$

$$= \frac{1 - e^{-(m+1)\beta}}{(m+1)(1 - e^{-\beta})} .$$

Substituting this result in (23) yields

$$\alpha^2 = \sum_{m=0}^{P(n)} \frac{n_m(n)}{(m+1)Z(n)} (1 + e^{-\beta} + e^{-2\beta} + \ldots e^{-m\beta}). \tag{26}$$

Note that for the normal distribution (5) with asymptotic behaviour according to (6) and (7), the i-th moment becomes asymptotically equal to μ_0^i. Expanding the exponential in the numerator and denominator, (23) produces series expansions which are easily summed, so that for $n \to \infty$,

$$\alpha^2 = \frac{e^{\mu_0(n)\beta} - 1}{\mu_0(n)\beta} \tag{27}$$

$$= \frac{e^{\rho n\beta} - 1}{\rho n\beta} \tag{28}$$

This resummation of the series is uniformly convergent in β.

By reasonable adjustment of the relationship between the interaction potential, temperature and molecular weight (namely, $\beta \sim n^{-\frac{1}{2}}$), it was possible to fit existing data for the expansion factor of polystyrene chains (see Gordon et al. [6]). However, there are obvious shortcomings in the first approximation estimate of $\overline{R_m^2}$, as regards its dependence on the chain length n. Indeed, (21) conflicts in different ways with asymptotic behaviour at low and high m.

5. SECOND APPROXIMATION

As pointed out earlier, each set of m contacts should be weighted directly by the number of distinct lattice walks (chain configurations) in which it occurs. (20) and (21) approximate this prescription well for fixed $n \gg 1$ and $m \geq n^{\frac{1}{2}}$, by first selecting at random the beads to be paired, and then weighting the relative chances of each pairing scheme according to the number of distinct walks in which it occurs. Simple statistical arguments prove the following asymptotic relation for $\overline{R_m^2} / \overline{R_0^2}$:

$$\overline{R_m^2} / \overline{R_0^2} = 1 - cmn^{-\frac{1}{2}}, \quad c = \text{constant}, \quad n \to \infty, \quad m < \gamma, \tag{29}$$

and there are strong suggestions that $\gamma = (\text{const}) \times n^{\frac{1}{4}}$. (For $\overline{R_1^2}$, the specified cycle has mean size $(\text{const}) \times n^{\frac{1}{2}}$, with standard deviation $(\text{const}) \times n^{\frac{3}{4}}$, as $n \to \infty$. If there are $\gamma > (\text{const}) \times n^{\frac{1}{4}}$ specified cycles, they become so closely spaced that their effects are no longer independent, and the linearity in (29) is lost. (5) shows that for $n \to \infty$, $\overline{r^2_\mu} \to \overline{r^2}$; that is, almost all walks have $m \approx \mu$ contacts. But $\overline{r^2} \equiv \overline{R^2}$ and accordingly, refinement of our theory depends on the sequence $\overline{R^2_0}$, $\overline{R^2_1}$, $\overline{R^2_2}$, which justifies the idea behind the classical perturbation treatment (see Fixman [4]).

The following model equation is proposed and tested below

$$\overline{R^2_m} / \overline{R^2_0} = (1 + KQ(1 - (1 - Q^{-1})^m))/(m + 1), \tag{30}$$

where

$$K = (2\overline{R^2_1} - \overline{R^2_0}) / \overline{R^2_0}, \tag{31}$$

and

$$Q = (2\overline{R^2_1} - \overline{R^2_0}) / (4\overline{R^2_1} - \overline{R^2_0} - 3\overline{R^2_2}). \tag{32}$$

If $\overline{R^2_1}$ and $\overline{R^2_2}$ are assumed to obey (29), as they should, then (30)-(32) ensure that (29) is obeyed for all m (closure relation). Besides, the second-approximation (30) quantifies, for our model, the first approximation inequality (21) (valid at high m and fixed n), since

$$\overline{R^2_m} / \overline{R^2_0} \sim Q/(m + 1), \quad \text{for } m \gg 1. \tag{33}$$

Combining (30) with (16) we can obtain a partially summed equation for α^2:

$$\alpha^2 = \sum_{m=0}^{P(n)} n_m \left[QK(f(y,m) - f(y^*,m)) + f(y,m) \right] / Z(n), \tag{34}$$

where

$$f(y,m) = [(1 + y)^{m+1} - 1] / y(m + 1), \tag{35}$$

$$y = e^{-\beta} - 1, \tag{36}$$

and

$$y^* = y(1 - Q^{-1}). \tag{37}$$

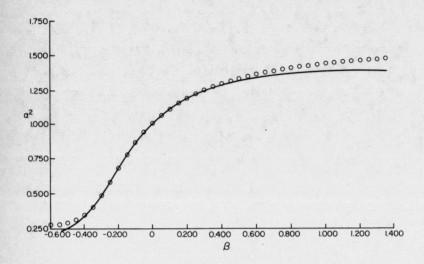

Figure 3

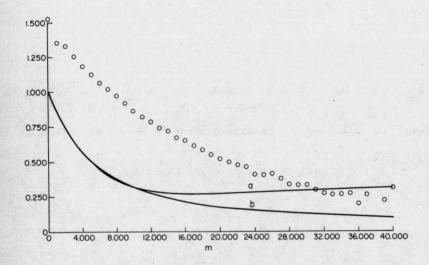

Figure 4

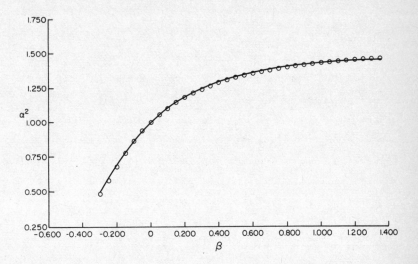

Figure 5

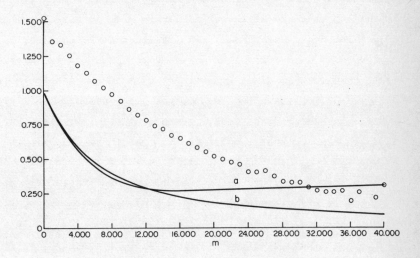

Figure 6

(34) has the advantage over (28) that β becomes independent of n, and the effect of the previous dependence $\beta \sim n^{-\frac{1}{2}}$ is recovered from $\overline{R_m^2}$ - that is, from the graph theory.

Equation (34), obtained from asymptotic considerations (n → ∞) is now tested against new Monte Carlo experiments for 3-choice diamond-lattice walks of n = 100 steps. All the quantities in (34) were obtained as averages from 10^5 such random walks. Figure 3 shows that the fit of (34) is very close for the range $-0.4 < \beta < +0.4$. The underlying $\overline{R_m^2}$ and $\overline{r_m^2}$ behaviour (see (30)) is shown in Figure 4. Figure 5 is a much improved fit resulting from letting the parameters $\overline{R_1^2}$ and $\overline{R_2^2}$ in (30) float for least-square optimisation, rather than taking their mean Monte Carlo values. Again, Figure 6 gives the underlying $\overline{r_m^2}$ behaviour (unchanged) and the $\overline{R_m^2}$ behaviour for this case. All deviations between Monte Carlo data and the fitted equations are due to three causes: (i) convergence of the Monte Carlo results requires at least 10^5 walks; (ii) convergence of the chain length requires at least 100 steps; (iii) inaccuracy of the model which (30) assumed. As regards the deviations between plots (a) and (b) in Figures 4 and 6 for large m, these are essentially due to (i), since it is an algebraic consequence of (8) that for the highest m observed in a sample of walks, $\overline{r_m^2} = \overline{R_m^2}$, as is apparent. (The highest m would increase on increasing the sample size.) The improvement in fit in passing from Figure 3 to Figure 5 results from adjustments in $\overline{R_1^2}$ and $\overline{R_2^2}$ by less than 0.1% and 1% respectively from their observed (Monte Carlo) values. In the practically important range of α^2, these minute adjustments almost cure the discrepancies arising from (i), (ii) and (iii), and that encourages the hope that (iii) alone (that is, the inadequacy of the model assumption) is of little significance. It has also been checked that the fit is little changed by replacing the Monte-Carlo data for n_m by an appropriate Gaussian form - that is, the form approached as n → ∞. It is concluded that the 'finite-chain' effect (lack of asymptoticity in n → ∞) is well modelled by (34), which is important for applications in polymer science. The model is also well suited to introduce the usual chain-stiffness effects arising from short-range inter-actions, as will be demonstrated elsewhere.

REFERENCES

1. L. Comtet, Advanced Combinatorics, D. Reidel Publishing Co. Dordrecht / Boston, Mass., 1974.

2. C. Domb and G. J. Joyce, Cluster expansion for a polymer chain, J. Phys. (C) $\underline{5}$ (1972), 956-976.

3. S. F. Edwards and K. F. Freed, Cross linkage problem of polymers: I - The method of second quantization applied to the cross linkage problem of polymers, J. Phys. (C) $\underline{3}$ (1970), 739-750.

4. M. Fixman, Excluded volume in polymer chains, J. Chem. Phys. $\underline{23}$ (1955), 1656-1659.

5. L. Flatto, The multiple range of two dimensional recurrent walks, Ann. Prob. $\underline{4}$ (1976), 228-248.

6. M. Gordon, J. A. Torkington and S. B. Ross-Murphy, The graph-like state of matter. 9. Statistical thermodynamics of dilute polymer solutions, Macromol. $\underline{10}$ (1977), 1090-1100.

7. N. C. Jain and W. E. Pruitt, The range of transient random walk, J. Analyse Math. $\underline{24}$ (1971), 369-393.

8. C. J. Preston, Gibbs states on countable sets, Cambridge University Press, Cambridge, 1974.

9. H. Yamakawa, Modern theory of polymer solutions, Harper and Row, New York, 1971.

M. Gordon and J. A. Torkington
Department of Chemistry
University of Essex
Wivenhoe Park
Colchester CO4 3SQ
England

Michel Las Vergnas
On Eulerian partitions of graphs

1. INTRODUCTION

The present paper is a brief survey of some different results of an algebraic nature concerning Eulerian partitions of graphs. Quoted theorems are mainly numerical equalities or inequalities, or polynomial identities, involving the numbers of Eulerian partitions with given numbers of circuits of graphs. Several of these results relate the numbers of Eulerian partitions (with and without crossings) of graphs imbedded in the sphere, the projective plane or the torus to certain evaluations of Tutte polynomials.

Graphs considered in the paper are finite with possibly loops and multiple edges. A loop incident to a vertex x contributes 2 to the degree of x in the undirected case, and contributes 1 to both the indegree and outdegree of x in the directed case.

A <u>circuit</u> of an undirected graph is a closed tour using every edge at most once, considered up to its initial (= terminal) vertex and direction. We consider that there are two ways of travelling along a loop (as along any other edge). In this respect it is convenient to divide every loop into two <u>half-loops</u> by a dummy vertex. We describe a circuit as a sequence of vertices, non-loop edges and half-loops. This convention is justified by topological applications in Sections 3, 4, 5.

In the directed case, a <u>directed circuit</u> is a closed tour using every edge at most once, consistently with direction of edges, and considered up to its initial (= terminal) vertex. Loops are considered as directed: there is only one way to travel along a loop in a directed graph.

An <u>Eulerian partition</u> of an undirected (resp. a directed) graph H is a set of circuits (resp. directed circuits) partitioning the edge-set of H. We say that an undirected (resp. a directed) graph H is <u>Eulerian</u> if H has an <u>Eulerian circuit</u> - that is, an Eulerian partition consisting of a unique circuit. As is well known, an undirected (resp. a directed) graph H is Eulerian if and only if H is connected and $d_H(x)$ is even (resp.

$d_H^-(x) = d_H^+(x))$ for every vertex x.

We denote by $f_k(H)$ the number of Eulerian partitions of H consisting of k circuits.

2. THE MARTIN POLYNOMIAL OF AN EULERIAN GRAPH

Let H be an Eulerian directed graph. We define the <u>Martin polynomial</u> m(H) of H by

$$m(H;\zeta) = \sum_{k \geq 0} f_{k+1}(H)(\zeta - 1)^k.$$

The following theorem may be found in [6]:

<u>Theorem 2.1</u>. The Martin polynomial of an Eulerian directed graph H is a polynomial with non-negative coefficients. Equivalently, the following inequalities hold:

$$\sum_{i \geq 0} (-1)^i \binom{k+i}{i} f_{k+i+1}(H) \geq 0 \quad \text{for} \quad k = 1, 2, \ldots . \quad \square$$

The proof of Theorem 2.1 uses three lemmas:

<u>Lemma 2.2</u>. We have

$$m(\vec{B}_n;\zeta) = \zeta(\zeta + 1) \ldots (\zeta + n - 2),$$

where $\vec{B}_n$ denotes the directed graph consisting of n loops incident to a unique vertex, $n \geq 2$.

<u>Proof</u>. Clearly $f_k(\vec{B}_n)$ is equal to the number of permutations of n elements with k cycles. Hence $f_k(\vec{B}_n) = c(n,k)$ the signless Stirling number of the first kind (defined by the identity $\zeta(\zeta + 1) \ldots (\zeta + n - 1) = \sum_{k \geq 0} c(n,k)\zeta^k$). The result follows. $\square$

Consider an Eulerian directed graph H with vertex-set V. Let x be a vertex of H, and σ be a bijection from the set of non-loop edges with terminal vertex x onto the set of non-loop edges with initial vertex x. We denote by H_σ the graph with vertex-set V\{x} obtained from H by the transformation shown in Figure 1. The bijection σ is given by

$e_i' \longrightarrow e_i$ ($i = 1, 2, \ldots, d - n$), where $d = d_H^-(x) = d_H^+(x)$ and n is the number of loops incident to x (in Figure 1, d = 4 and n = 1).

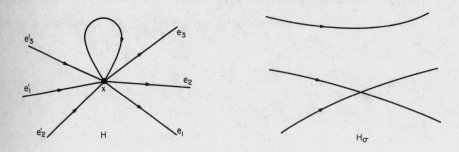

Figure 1

We say that H_σ is a graph <u>derived from H at x</u>. There are $(d - n)!$ possible bijections σ, and hence there are $(d - n)!$ graphs derived from H at x.

Lemma 2.3. Suppose $x \in V$ is not a cut-vertex of H, and that $d \geq 2$ and $n \leq d - 1$. We have

$$f_k(H) = \sum_{H'} \sum_{0 \leq j \leq \min(n,k-1)} \sum_{j \leq i \leq n} \binom{n}{i} \frac{(d - i - 1)!}{(d - n - 1)!} c(i,j) f_{k-j}(H'),$$

where the first sum is taken over the $(d - n)!$ graphs H' derived from H at x. □

From Lemma 2.3 follows

Lemma 2.4. Suppose $x \in V$ is not a cut-vertex of H, and that $d \geq 2$ and $n \leq d - 1$. We have

$$m(H;\zeta) = \left(\prod_{2 \leq i \leq n+1} (\zeta + d - i)\right) \sum_{H'} m(H';\zeta),$$

where the sum is taken over the $(d - n)!$ graphs H' derived from H at x. □

The polynomial $m(H)$ was introduced by P. Martin in [7] for 4-valent Eulerian directed graphs. The definition of $m(H)$ in [7] is by induction on the number of vertices, using Lemma 2.4 with $d = 2$ as the inductive step. The existence of $m(H)$ in this case, and the relations $f_k(H) = \sum \binom{i}{k-1} m_i(H)$ $(i \geq k - 1)$, where the coefficients $m_i(H)$ are defined by the identity $m(H;\zeta) = \sum m_i(H) \zeta^i$ $(i \geq 0)$, equivalent to our definition of $m(H)$, are obtained as theorems in [7].

Theorem 2.1 solves a conjecture stated in [7, p.126]. We mention that Lemma 2.4 can be generalized to cut-vertices, although the formula is more complicated (see [6]). In the case $d = 2$, suppose that x is a cut-vertex of H; then one of the two derived graphs has exactly two connected components H_1, H_2 (the other being connected), and we have $m(H;\zeta) = m(H_1;\zeta) m(H_2;\zeta)$ (see [7]).

The following result may be found in [6]:

Proposition 2.5. Let H be an Eulerian directed graph with minimum degree $2d \geq 4$. Then $m(H;\zeta)$ is divisible by $m(\vec{B}_d;\zeta)$. Equivalently, the following relations hold:

$$m(H;-i) = \sum_{k \geq 0} (-i - 1)^k f_{k+1}(H) = 0, \text{ for } i = 0, 1, \ldots, d - 2. \quad \square$$

Note that the total number of Eulerian partitions is $\Pi(d_H^+(x))!$ $(x \in V)$. Hence $m(H;2) = \Pi (d_H^+(x))!$ $(x \in V)$.

Similar results hold in the undirected case, the only difference being the presence of powers of 2 in these formulas (see [6]). Let H be an Eulerian (undirected) graph. Then $m(H;\zeta) = \sum f_{k+1}(H)(\zeta - 2)^k$ $(k \geq 0)$ is a polynomial with non-negative coefficients. We have $f_k(B_n) = 2^{n-k} c(n,k)$, and hence $m(B_n;\zeta) = \zeta(\zeta + 2) \ldots (\zeta + 2n - 4)$. The formula in Lemma 2.4 is

$$m(H;\zeta) = (\prod_{2 \leq i \leq n+1} (\zeta + 2d - 2i)) \sum_{H'} m(H';\zeta),$$

where the sum is taken over the $(2d - 2n - 1)!! = \dfrac{(2d - 2n)!}{2^{d-n}(d-n)!}$ derived graphs of H at x (in the undirected case, the derived graphs are in one-one correspondence with partitions into unordered pairs of the set of 2d - 2n non-loop edges incident at x). The polynomials m(H) in the undirected case are introduced in [7] for 4-valent and 6-valent graphs.

3. <u>EULERIAN PARTITIONS WITHOUT CROSSINGS OF 4-VALENT GRAPHS IMBEDDED IN THE PLANE</u>

Let H be a connected 4-valent (undirected) graph imbedded in the plane. As is well known, the faces of H can be colored in two colors such that any two faces with a common edge have different colors. The graph H has thus white faces and black faces (say). Consider a vertex x of H: the four non-loop edges and half-loops incident at x, or <u>incidences</u> at x, are cyclically ordered by the imbedding. There are three partitions into two (unordered) pairs of the four incidences at x: with the notation of Figure 2 we call $\{e_1, e_2\} \{e_3, e_4\}$ a <u>white non-crossing</u> transition, $\{e_1, e_4\} \{e_2, e_3\}$ a <u>black non-crossing transition</u>, and $\{e_1, e_3\} \{e_2, e_4\}$ a <u>crossing transition</u>.

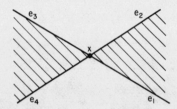

Figure 2

Observe that an Eulerian partition determines a transition at every vertex and conversely - we let $\gamma = (\ldots, e_1, x, e_2, \ldots)$ and $\gamma' = (\ldots, e_3, x, e_4, \ldots)$ be the two occurrences of x in circuits of an Eulerian partition: the corresponding transition at x is then $\{e_1, e_2\} \{e_3, e_4\}$. We say that a vertex x is a <u>white</u> (resp. <u>black</u>) <u>non-</u>

crossing (crossing) of a given Eulerian partition if the corresponding transition at x is a white (black) non-crossing transition (resp. a crossing transition).

We denote by $f_k^x(H)$ the number of Eulerian partitions without crossings of H. Eulerian circuits without crossings have been considered by several authors: W. T. Tutte and C. A. B. Smith have calculated $f_1^x(H)$ by means of determinants [12]. Their result is related to a natural one-one correspondence established by A. Kotzig between Eulerian circuits without crossings and spanning trees of the graph of white faces of H (see [2, Theorem 14]).

We recall that the <u>graph of white faces</u> of H is a graph G imbedded in the plane, defined as follows:

(1) the vertices of G are in one-one correspondence with the white faces of H; every white face F of H contains exactly one vertex v(F) of G;

(2) the edges of G are in one-one correspondence with the vertices of H; if x is a vertex of H, and F_1, F_2 are the two (possibly equal) white faces of H incident to x, then the edge e(x) of G corresponding to x is contained in $F_1 \cup F_2 \cup \{x\}$, contains x, and joins $v(F_1)$ and $v(F_2)$.

As is easily seen, a graph G with these properties is topologically unique (i.e. unique up to homeomorphism). Clearly G and G*, the <u>graph of black faces</u> of H, are dually imbedded in the plane.

The above construction can be reversed. Given any graph G imbedded in the plane, there is a topologically unique 4-valent graph H imbedded in the plane such that G is the graph of white faces and G* the graph of black faces of H. The graph H is called the <u>medial</u> graph of G.

We have generalized the result of Kotzig to any Eulerian partition without crossings in the following theorem [5]:

Theorem 3.1. Let H be a connected 4-valent graph imbedded in the plane. Then the number of circuits of an Eulerian partition without crossings of H is equal to

$$|X| + r(\underline{C}(G)) - 2r_{\underline{C}(G)}(e(X)) + 1 = |V\backslash X| + r(\underline{C}(G^*)) - 2r_{\underline{C}(G^*)}(e^*(V\backslash X)) + 1$$

$$= |V| - r_{\underline{C}(G)}(e(X)) - r_{\underline{C}(G^*)}(e^*(V\backslash X)) + 1,$$

where V denotes the vertex-set of H, X the set of white transitions of the partition, G (resp. G*) the graph of white (resp. black) faces of H, $\underline{C}(G)$ the circuit-geometry (circuit-matroid) of G, $r(\underline{C}(G))$ the rank of $\underline{C}(G)$, and $r_{\underline{C}(G)}(e(X))$ the rank in $\underline{C}(G)$ of $e(X)$, the set of edges of G corresponding to X. □

Define the Martin polynomial of the graph H imbedded in the plane by

$$m^x(H;\zeta) = \sum_{k\geq 0} f^x_{k+1}(H)(\zeta - 1)^k.$$

We denote by $t(G)$ the Tutte polynomial of G (also called the 'dichromatic polynomial'). We recall that the Tutte polynomial $t(M)$ of a combinatorial geometry (matroid) M on a finite set E is a polynomial in two variables, defined by

$$t(M;\zeta,\eta) = \sum_{X\subseteq E} (\zeta - 1)^{r(M)-r_M(X)} (\eta - 1)^{|X|-r_M(X)};$$

$t(G)$ can be defined as $t(G) = t(\underline{C}(G))$ (see [13]).
The following result may be found in [7]:

<u>Corollary 3.2.</u> Let G be a graph imbedded in the plane, and let H be the medial graph of G. Then

$$t(G;\zeta,\zeta) = m^x(H;\zeta). \quad □$$

Actually in [7], the identity $t(G;\zeta,\zeta) = m^x(H;\zeta)$ is taken as the definition of $m^x(H)$. Inductive proofs show that $f^x_k(H) = \sum_{i\geq k-1} \binom{i}{k-1} m^x_i(H)$ (see [7, pp.67-68]).

As is well known, edges of a connected 4-valent graph H imbedded in the plane can be directed in such a way that the border of every face is a directed circuit. We denote by $\vec{H}$ the directed graph obtained, unique up to the reversal of all edge directions. Clearly an Eulerian partition of H without crossings corresponds to an Eulerian partition of $\vec{H}$, and conversely. Hence $m^x(H;\zeta) = m(\vec{H};\zeta)$ [7].

Applying Edmonds' Intersection Theorem, we get from Theorem 3.1 the following max-min theorem (see [5]):

<u>Corollary 3.3.</u> Let H be a connected 4-valent graph imbedded in the plane. Then the maximum number of circuits of an Eulerian partition without crossings of H is one more than the minimum cardinality of a set of vertices connecting all white faces and connecting all black faces. □

A set of vertices X of H connects all white faces and connects all black faces if the two sets $X \cup F_W$ and $X \cup F_B$ are connected parts of the plane, where F_W (resp. F_B) denotes the union of all white (resp. black) faces of H. In fact, in Corollary 3.3 the maximum can be taken over all Eulerian partitions of H since we have [5]:

<u>Proposition 3.4.</u> The maximum number of circuits partitioning the edge-set of an Eulerian graph imbedded in the plane is equal to the maximum number of circuits of an Eulerian partition without crossings.

4. <u>CROSSING CIRCUITS</u>

Let H be a 4-valent graph. Suppose that at every vertex the three transitions are colored in the three colors black, white and red in a one-one manner. Given a 3-partition $V = X + Y + Z$ of the vertex-set V of H, we denote by $f_H(X,Y,Z)$ the number of circuits of the Eulerian partition of H with X, Y and Z as sets of white, black and red transitions, respectively. We set $g_H(X,Y,Z) = (-1)^{|Z|}(-2)^{f_H(X,Y,Z)}$.

<u>Lemma 4.1.</u> $g_H(X,Y,Z) = \sum_{A \subseteq Z} g_H(X \cup A, V \setminus (X \cup A), \emptyset)$. □

Lemma 4.1 is stated in [5]. However the ideas are implicit in [7], [8] and [10].

<u>Applications:</u>

(1) Suppose H is a 4-valent Eulerian directed graph. At each vertex x we color black and white the two transitions $\{e_1', e_1\}\{e_2', e_2\}$ and $\{e_1', e_2\}\{e_2', e_1\}$ and red the transition $\{e_1', e_1\}\{e_2', e_2\}$ (see Figure 3):

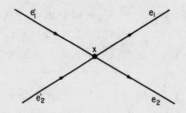

Figure 3

Applying Lemma 4.1 with $X = Y = \emptyset$, we get (see [7]):

<u>Proposition 4.2</u>. Let H be a 4-valent Eulerian directed graph. Then

$$m(H;-1) = \sum_{k \geq 0} (-2)^k f_{k+1}(H) = (-1)^{\alpha_0(H)} (-2)^{c(H)-1},$$

where $\alpha_0(H)$ is the number of vertices of H, and $c(H)$ is the number of anti-directed circuits of H. □

(2) Suppose H is a 4-valent graph imbedded in a surface. At each vertex we color black and white the two non-crossing transitions and red the crossing transition. We say that circuits of the unique Eulerian partition of H crossing at every vertex are <u>crossing circuits of H</u>.

Applying Lemma 4.1 with $X = Y = \emptyset$, we get (see [5]):

<u>Proposition 4.3</u>. Let H be a 4-valent graph imbedded in a surface. Then

$$\sum_{k \geq 1} (-2)^k f_k(H) = (-1)^{\alpha_0(H)} (-2)^{c(H)},$$

where $f_k(H)$ is the number of Eulerian partitions of H without crossings and consisting of k circuits, $\alpha_0(H)$ is the number of vertices of H, and $c(H)$ is the number of crossing circuits of H. □

Note that when the surface is orientable and the faces of H can be 2-colored, Proposition 4.3 is a consequence of Proposition 4.2, by an extension of the construction given in Section 3. It follows from Proposition 4.3 that H has exactly one crossing circuit if and only if the number of Eulerian circuits without circuits is odd.

When the considered surface is the plane, one obtains from Proposition 4.3 and Corollary 3.2 a result due to P. Martin [8] and to P. Rosenstiehl and R. C. Read [10]:

<u>Corollary 4.4</u>. Let G be a graph imbedded in the plane. Then

$$t(G;-1,-1) = (-1)^{\alpha_1(G)}(-2)^{c(H)-1},$$

where $\alpha_1(G)$ is the number of edges of G, and $c(H)$ is the number of crossing circuits of the medial graph H of G. □

Since $t(G;-1,1)$ and $t(G;1,1)$ (the number of spanning trees of G) have equal parities, Corollary 4.4 contains the following result of H. Shank: a connected 4-valent graph H imbedded in the plane has exactly one crossing circuit if and only if the number of spanning trees of the graph of white faces of H is odd (see [11, Theorem 4]).

We mention the following related problem on crossing circuits in the plane, due to Gauss: given a cyclic word γ in an alphabet V such that every letter occurs twice, does there exist a 4-valent graph with vertex-set V imbedded in the plane such that γ is the unique crossing circuit of H? We refer to [9] and [10] for an algebraic solution of this problem, and a bibliography of different other solutions.

5. EULERIAN PARTITIONS OF 4-VALENT GRAPHS IMBEDDED IN SURFACES

A natural question arises whether results of the preceding sections for the plane (or, equivalently, the sphere) can be generalized to other surfaces. We have given such generalizations in [5] for the projective plane and the torus.

We recall that an imbedding of a graph H in a surface S is <u>cellular</u> if every face of H is homeomorphic to an open 2-disc. If the graph H is cellularly imbedded in S, we have $\alpha_0(H) - \alpha_1(H) + \alpha_2(H) = \chi(S)$ and conversely, where $\alpha_0(H)$ (resp. $\alpha_1(H), \alpha_2(H)$) is the number of vertices (resp. edges, faces) of H, and $\chi(S)$ is the Euler characteristic of S ($\chi(S) = 2$ for the sphere, 1 for the projective plane, and 0 for the torus). Note that a graph cellularly imbedded in a surface is necessarily connected.

Let H be a 4-valent graph cellularly imbedded in a surface S (orientable or not) in such a way that its faces can be 2-colored. The graph G of white faces of H is defined as in Section 3. It can be easily shown that G is cellularly imbedded in S. Conversely, a graph G cellularly imbedded in S has a topologically unique medial. The graph G* of black faces of H is a topological dual of G in S. White (black) non-crossings and crossings of an Eulerian partition are defined as in the plane case. We then have the following result [5]:

__Theorem 5.1.__ Let H be a 4-valent graph cellularly imbedded in the sphere, the projective plane or the torus, in such a way that its faces can be 2-colored. Then the number of circuits of an Eulerian partition of H without crossings and with X as its set of non-crossings, is given by
$$\min\{|X| + r(\underline{C}(G)) - 2r_{\underline{C}(G)}(e(X)) + 1, |V\setminus X| + r(\underline{C}(G^*)) - 2r_{\underline{C}(G^*)}(e^*(V\setminus X)) + 1\},$$
where V is the vertex-set of H, and G (resp. G*) is its graph of white (resp. black) faces. □

By a theorem of J. Edmonds ([1, Theorem 1]), up to the natural bijection between edges of G and G*, any circuit of the binary geometry $\underline{B}(G^*) = (\underline{C}(G^*))^{\perp}$ (the __bond-geometry__ of G*) is a disjoint union of circuits of $\underline{C}(G)$. In particular, the Tutte polynomial of the pair $\underline{B}(G^*), \underline{C}(G)$ is defined (see [3], [4]). We have

$$t(\underline{B}(G^*), \underline{C}(G); \zeta, \eta, \xi) =$$
$$= \sum_{X \subseteq E} (\zeta-1)^{r(\underline{C}(G))-r_{\underline{C}(G)}(X)} (\eta-1)^{|X|-r_{\underline{B}(G^*)}(X)} \xi^{r(\underline{B}(G^*))-r(\underline{C}(G))-(r_{\underline{B}(G^*)}(X)-r_{\underline{C}(G)}(X))}$$

where E is the edge-set of G and we denote by the same letter X the set of edges corresponding to X in G*.

We call $t(\underline{B}(G^*), \underline{C}(G))$, abbreviated to $t(G, G^*)$, the __Tutte polynomial of the graph imbedding__ G. The degree of $t(G, G^*)$ in ξ is $r(\underline{B}(G^*)) - r(\underline{C}(G))$, and so this degree is $2 - \chi(S)$ when G is cellularly imbedded in the surface

S. Note that the polynomial $t(G,G^*)$ is the usual Tutte polynomial of G when S is the sphere.

<u>Corollary 5.2.</u> The following identities hold:
(<u>projective plane</u>)

$$t(G,G^*;\zeta,\zeta,1)) = \frac{1}{\zeta}(t(G;\zeta,\zeta) + t(G^*;\zeta,\zeta)) = m^\times(H;\zeta),$$

(<u>torus</u>)

$$t_2(G,G^*;\zeta,\zeta) + (\zeta - 1)t_1(G,G^*;\zeta,\zeta) + t_0(G,G^*;\zeta,\zeta)$$
$$= t(G;\zeta,\zeta) - (\zeta - 1)t(L;\zeta,\zeta) + t(G^*;\zeta,\zeta) = m^\times(H;\zeta),$$

where $t_0(G,G^*)$, $t_1(G,G^*)$ and $t_2(G,G^*)$ are defined by the identity

$$t(G,G^*:\zeta,\eta,\xi) = \xi^2 t_2(G,G^*:\zeta,\eta) + \xi t_1(G,G^*;\zeta,\eta) + t_0(G,G^*:\zeta,\eta),$$

and L is the Higgs lift of $\underline{C}(G)$ in $\underline{B}(G^*)$. □

From Proposition 4.3 and Corollary 5.2 we get:

<u>Corollary 5.3.</u> Let G be a graph cellularly imbedded in the projective plane or the torus, with dual graph G^*. Then
(<u>projective plane</u>)

$$t(G,G^*;-1,-1,1) = -t(G;-1,-1) - t(G^*;-1,-1) = (-1)^{\alpha_1(G)}(-2)^{c(H)-1},$$

(<u>torus</u>)

$$t_2(G,G^*;-1,-1) - 2t_1(G,G^*:-1,-1) + t_0(G,G^*;-1,-1) = (-1)^{\alpha_1(G)}(-2)^{c(H)-1},$$

where $c(H)$ is the number of crossing circuits of the medial graph H of G. □

It follows easily from Corollary 5.3 that the medial graph of a graph cellularly imbedded in the projective plane or in the torus has a **unique** crossing circuit if and only if the numbers of spanning trees of G and G^*

are of different parities. A max-min theorem analogous to Corollary 3.3 can be obtained from Theorem 5.1 for the projective plane and the torus [5]; however, Proposition 3.4 holds only for the sphere.

We mention that both the projective plane case and the torus case of Theorem 5.1 contain the sphere case: given a connected graph G imbedded in the sphere, it is an easy exercise to construct cellular imbeddings in the projective plane or the torus by adding one or two loops at some vertex and to get Theorem 5.1 for G from Theorem 5.1 applied to these imbeddings. However, the proof of Theorem 5.1 for the projective plane and the torus uses Theorem 5.1 for the sphere.

Theorem 5.1 holds only for the sphere, the projective plane and the torus. Moreover it turns out that, except in these three cases, $\underline{C}(G)$ and $\underline{C}(G^*)$ are not sufficient to determine the number of circuits of an Eulerian partition without crossings from its set of white non-crossings. An example is given by the graph imbedding in the Klein bottle of Figure 4 (the Klein bottle is represented by a rectangle with edges identified as indicated by letters and arrows): x_1 and x_2 cannot be distinguished within $\underline{C}(G)$ and $\underline{C}(G^*)$ both of rank 0; however, the numbers of circuits of the two Eulerian partitions without crossings having sets of white non-crossings $\{x_1\}$ and $\{x_2\}$ are different (respectively, 1 and 2). Analogous counterexamples can be constructed for surfaces of negative characteristic.

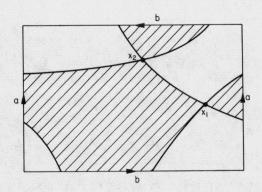

Figure 4

The above example implies that to extend Theorem 5.1 further, other algebraic invariants besides $\underline{C}(G)$ and $\underline{C}(G^*)$ have to be considered. The nature of these invariants is an open problem.

REFERENCES

1. J. Edmonds, On the surface duality of linear graphs, J. Res. Nat. Bur. Standards $\underline{69}$ B (1965), 121-123.

2. A. Kotzig, Eulerian lines in finite 4-valent graphs and their transformations, Proc. Colloq. Tihany 1966, Hungarian Acad. Sci. Budapest, North-Holland, Amsterdam (1968), 219-230.

3. M. Las Vergnas, Extensions normales d'un matroide, polynôme de Tutte d'un morphisme, C.R. Acad. Sci. Paris (A) $\underline{280}$ (1975), 1479-1482.

4. M. Las Vergnas, The Tutte polynomial of a morphism of combinatorial geometries I, J. Combinatorial Theory (A) (to appear).

5. M. Las Vergnas, Eulerian circuits of 4-valent graphs imbedded in surfaces, Proc. Colloq. on Algebraic Methods in Graph Theory, Szeged 1978 (to appear).

6. M. Las Vergnas, The Martin polynomial of an Eulerian graph (to appear).

7. P. Martin, Enumérations eulériennes dans les multigraphes et invariants de Tutte-Grothendieck, Thesis, Grenoble, 1977.

8. P. Martin, Remarkable valuation of the dichromatic polynomial of planar multigraphs, J. Combinatorial Theory (B) $\underline{24}$ (1978), 318-324.

9. P. Rosenstiehl, Solution algébrique du problème de Gauss sur las permutation des points d'intersection d'une ou plusieurs courbes fermées du plan, C.R. Acad. Sci. Paris (A) $\underline{283}$ (1976), 551-553.

10. P. Rosenstiehl and R. C. Read, On the principal edge-tripartition of a graph, Discrete Math. $\underline{3}$ (1978), 195-226.

11. H. Shank, The theory of left-right paths, Combinatorial Math. III, Lecture Notes in Math. $\underline{452}$, Springer-Verlag, Berlin (1975), 42-54.

12. W. T. Tutte and C. A. B. Smith, On unicursal paths in a network of degree 4, Amer. Math. Monthly $\underline{4}$ (1941), 233-237.

13. D. J. A. Welsh, Matroid Theory, Academic Press, London, 1976.

M. Las Vergnas
Centre National de la Recherche Scientifique
Université Pierre et Marie Curie (U.E.R. 48)
4 Place Jussieu
75230 Paris, France

Colin McDiarmid
Colouring random graphs badly

1. INTRODUCTION

We shall consider how badly graph colouring heuristics may perform on the average. We do this by investigating the behaviour for large random graphs of two graph functions (the achromatic number ψ and the canonical achromatic number ψ_c), and we also give a result which more directly concerns colouring heuristics.

Firstly, why do we want to colour graphs, and why resort to heuristics to do so? A (proper) <u>colouring</u> of a graph G is an assignment of colours to the vertices so that no two adjacent vertices have the same colour. The <u>chromatic number</u> $\chi(G)$ is the least number of colours used in any colouring of G.

Many practical problems can be modelled in terms of colouring a graph using few colours. A standard example involves the storage of chemicals, where certain pairs of chemicals are to be kept in separate compartments of a warehouse since they may interact, and we wish to use few compartments. We may construct a graph G with a vertex for each chemical and join a pair of vertices when the corresponding chemical interact. A colouring of G using k colours then corresponds to a storage scheme using k compartments. A second standard example involves timetabling examinations, where certain pairs of examinations may not occur in the same period since some student takes both, and we wish to use few periods.

It would be very useful to have a good algorithm for colouring a large graph G with (say) $\chi(G)$ colours. Unfortunately no such algorithm is known; and indeed it seems unlikely that any such algorithm will be found, since the problem of determining $\chi(G)$ is known to be NP-hard [1]. Thus if there were a good algorithm there would also be good algorithms for such notoriously difficult problems as determining whether a graph has a Hamiltonian cycle. Further the problem of determining $\chi(G)$ to within a factor $2 - \varepsilon$ is still NP-hard [4].

For large graphs then we must use colouring heuristics - that is, algorithms which produce proper colourings using (we hope) few colours.

An example of such a heuristic is the simple sequential colouring algorithm SA. Suppose that a graph G has vertices $v_1, v_2, v_3, \ldots$. Then SA colours the vertices of G sequentially with colours $c_1, c_2, \ldots$. We start by colouring v_1 with colour c_1, then we colour v_2 with colour c_1 unless v_1 and v_2 are adjacent in which case we use colour c_2, and in general when $v_1, \ldots, v_m$ have been coloured we colour v_{m+1} with the available colour c_i with least index. Note that SA will colour G with just $\chi(G)$ colours if the vertices are presented in an appropriate order.

How good are these heuristics? Denote by $A(G)$ the number of colours used on a graph G by algorithm A. Then $A(G)/\chi(G)$ is a measure of how well the algorithm A has performed on G. Let us define $\hat{A}(n)$ to be the maximum value of $A(G)/\chi(G)$ over all graphs G with n vertices. Of course $1 \leq \hat{A}(n) \leq n$. Johnson [7] showed that for many of the popular heuristics $\hat{A}(n)$ is of order n, and the best known growth rate for a (fast) heuristic is of order $n/\log n$. For example, by considering the graph in Figure 1, we may see that $\hat{SA}(2n) \geq \frac{1}{2}n$.

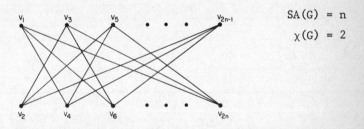

$SA(G) = n$

$\chi(G) = 2$

Figure 1

We have seen that colouring heuristics *can* behave very badly, but how often will we be so unlucky? We shall see that *usually* they behave quite well.

In order to talk about average behaviour we must choose some probability distributions, and we make here the simplest and most natural choice. Let p be fixed with $0 < p < 1$, and let $q = 1 - p$. We let G_n denote a random graph with vertex set $\{v_1, \ldots, v_n\}$ such that the $\binom{n}{2}$

possible edges occur independently with probability p. Thus if $p = \frac{1}{2}$, the probability that G_n has some property is just the proportion of the graphs on n (labelled) vertices with this property. (The assumption that the edges of G occur independently and with the same probability is of course unlikely to yield a good model of the sorts of graph that arise from (say) problems of chemical storage or of examination timetabling, but it is the obvious place to start our investigations.)

When the probability that the random graph G_n has some property tends to 1 as $n \to \infty$, we say that the property holds 'for almost all graphs G_n'. For example, we have the following results from [5]. (All logarithms are natural.)

For almost all graphs G_n,

$$\chi(G_n) \geq \frac{n \log(q^{-1})}{2 \log n}, \tag{1}$$

and

$$SA(G_n) \leq (1 + \varepsilon) \frac{n \log(q^{-1})}{\log n}, \tag{2}$$

and hence

$$\frac{SA(G_n)}{\chi(G_n)} \leq 2 + \varepsilon. \tag{3}$$

Thus the simple sequential colouring algorithm SA usually behaves quite well asymptotically. (This possibility was foreseen in [7].) In fact, (2) gives a good bound since for almost all graphs G_n

$$SA(G_n) \geq (1 - \varepsilon) \frac{n \log(q^{-1})}{\log n}. \tag{4}$$

We now define the two graphs functions which we shall use to give us information about how badly colouring heuristics may perform.

A colouring of a graph G with colours $c_1, \ldots, c_k$ is <u>complete</u> (or minimal) if for each pair of distinct colours c_i and c_j some vertex coloured c_i and some vertex coloured c_j are adjacent. The <u>achromatic number</u> $\psi(G)$ of G is the largest number of colours used in any complete colouring of

G. In the chemical storage problem the achromatic number would be the maximum number of compartments that might be needed when there is also a (compatible) requirement that certain chemicals are to be in the same compartment. All the usual colouring heuristics produce complete colourings, and in any case any colouring may easily be 'shrunk' to a complete colouring. Thus $\psi(G)$ is an upper bound on the number of colours we need ever use to colour G, and $\psi(G)/\chi(G)$ is a measure of how badly it is possible to colour G.

A colouring of a graph G with colours $c_1, \ldots, c_k$ is <u>canonical</u> if, for $1 \leq i < j \leq k$, each vertex coloured c_j is adjacent to some vertex coloured c_i (so that the vertices have been 'pushed as far left' as possible). The <u>canonical achromatic number</u> $\psi_c(G)$ is the largest number of colours used in any canonical colouring of G. Most sequential algorithms produce canonical colourings. Also, given any colouring, if we re-order the vertices so that vertices with the same colour come together and then apply SA we obtain a canonical colouring using no more colours. Thus $\psi_c(G)$ is an upper bound on the number of colours used on G by many sequential algorithms, and $\psi_c(G)/\chi(G)$ is a measure of how badly any such algorithm might perform on G.

We shall prove that, given $\varepsilon > 0$, for almost all graphs G_n

$$\frac{\psi(G_n)}{\chi(G_n)} \leq (2 + \varepsilon) \left| \frac{\log n}{\log(q^{-1})} \right|^{\frac{1}{2}}$$

and

$$\frac{\psi_c(G_n)}{\chi(G_n)} \leq 2 + \varepsilon .$$

Thus the usual behaviour of colouring heuristics is much better than the extreme behaviour. We shall also investigate 'sequential partitioning algorithms' and prove a result very similar to our result on canonical achromatic numbers.

2. RESULTS

Consider first the achromatic number $\psi(G)$.

Theorem 1. Given $\varepsilon > 0$, for almost all graphs G_n,

$$\frac{\psi(G_n)}{\chi(G_n)} \leq (2 + \varepsilon) \left| \frac{\log n}{\log(q^{-1})} \right|^{\frac{1}{2}}. \quad \square$$

This theorem follows immediately from (1), together with the following lemma (see [9]):

Lemma 2. Let $\varepsilon > 0$, and let

$$k = k(n) \sim (1 + \varepsilon) n \left| \frac{\log(q^{-1})}{\log n} \right|^{\frac{1}{2}}.$$

Then

$$P\{\psi(G_n) \geq k(n)\} \to 0 \quad \text{as } n \to \infty.$$

Proof. Let $\underset{\sim}{A} = (A_1, \ldots, A_k)$ be a partition of $\{v_1, \ldots, v_n\}$ into k set, of sizes $a_1, \ldots, a_k$, say. The probability that for each $i \neq j$ some vertex in A_i and some vertex in A_j are adjacent equals

$$\prod_{i<j} (1 - q^{a_i a_j}) \leq \exp\{-\sum_{i<j} q^{a_i a_j}\}$$

$$\leq \exp\{-\binom{k}{2} q^{n^2/k^2}\}.$$

But the number of partitions of $\{v_1, \ldots, v_n\}$ is certainly at most n^n. Hence

$$P\{\psi(G_n) \geq k\} \leq n^n \exp\{-\binom{k}{2} q^{n^2/k^2}\}$$

$$\leq \exp\{-n^{1+\eta+o(1)}\} \quad \text{for some } \eta > 0$$

$$\to 0 \quad \text{as } n \to \infty \quad \text{(fast)}. \quad \square$$

(Here and elsewhere we have omitted the arithmetic.)

Note that the above lemma really concerns the 'pseudo-achromatic' number of R. P. Gupta [6].

With rather more work we may show that the bound in Theorem 1 is of the right order. By considering a silly colouring heuristic we proved [9] that, given $\varepsilon > 0$, for almost all graphs G_n

$$\psi(G_n) \geq (2^{-\frac{1}{2}} - \varepsilon) n \left| \frac{\log(q^{-1})}{\log n} \right|^{\frac{1}{2}}, \dots$$

and hence, by (2),

$$\frac{\psi(G_n)}{\chi(G_n)} \geq (2^{-\frac{1}{2}} - \varepsilon) \left| \frac{\log n}{\log(q^{-1})} \right|^{\frac{1}{2}}.$$

Next we look at the canonical achromatic number $\psi_c(G)$.

Theorem 3. Given $\varepsilon > 0$, for almost all graphs G_n,

$$\frac{\psi_c(G_n)}{\chi(G_n)} \leq 2 + \varepsilon. \quad \square$$

This theorem follows immediately from (1), together with the following lemma. (We state the lemma in more detail than is necessary just to deduce Theorem 3.)

Lemma 4. Let $5 \leq v(n) < \log n / \log\log n$, and let

$$k(n) = \frac{n \log(q^{-1})}{\log n} \left(1 - v(n) \frac{\log\log n}{\log n} \right)^{-1}.$$

Then for n sufficiently large,

$$P\{\psi_c(G_n) \geq k(n)\} < \exp\{-n (\log n)^{v-4}\}.$$

Proof. Let $\underset{\sim}{A} = (A_1, \dots, A_m)$ be a partition of $\{v_1, \dots, v_n\}$ into $m \geq k$ sets. Let us say that $\underset{\sim}{A}$ is <u>canonical</u> for G_n if, for $i < j$, each vertex in A_j is adjacent to some vertex in A_i. Since the number of partitions of

$\{v_1, \ldots, v_n\}$ is certainly at most n^n (as before), it suffices to prove that (say)

$$P\{\underline{A} \text{ is canonical for } G_n\} < \exp\{-n (\log n)^{v-3+o(1)}\}. \tag{5}$$

Let $g = g(n) = n(\log n)^{-2}$, and for $i = 1, \ldots, m$, denote $|A_i|$ by a_i. Let t be the greatest integer i such that $a_1 + \ldots + a_i \leq n - g$. Note that $t \geq m - g \geq k - g$.

Now

$$P\{\underline{A} \text{ is canonical for } G_n\}$$
$$= (1-q^{a_1})^{n-a_1}(1-q^{a_2})^{n-(a_1+a_2)} \ldots (1-q^{a_{m-1}})^{a_m}$$
$$\leq \exp\{-(n-a_1)q^{a_1} - (n-(a_1+a_2))q^{a_2} - \ldots - (n-(a_1+\ldots+a_t))q^{a_t}\}$$
$$\leq \exp\{-g(q^{a_1} + \ldots + q^{a_t})\}$$
$$\leq \exp\{-gt\, q^{(n-g)/t}\}.$$

But

$$gt\, q^{(n-g)/t} \geq g(k-g)q^{n/(k-g)} \geq n(\log n)^{v-3+o(1)},$$

which yields (5) as required. □

Note that Lemma 4 really concerns the 'canonical pseudo achromatic number' of a graph G - that is, the maximum number of sets in a canonical partition of G.

For any graph G, of course $SA(G) \leq \psi_c(G)$. By Lemma 4 and (4) we have that, given any $\varepsilon > 0$, for almost all graphs G_n

$$\psi_c(G_n) < (1 + \varepsilon) SA(G_n). \tag{6}$$

Of course, (6) together with (3) gives us Theorem 3 again.

In this last part we consider a result which deals more directly with colouring heuristics, but which is closely related to the result on canonical achromatic numbers.

Many popular colouring heuristics are modifications of the simple sequential colouring algorithm SA (see for example [3], [7], [8]). The modifications may involve re-ordering the vertices or using 'bi-chromatic interchanges' (and we might not end up with a canonical colouring). They are intuitively reasonable, but might they make matters worse?

A graph algorithm is a <u>sequential partitioning algorithm</u> if it acts as follows on a graph G_n. First we put v_1 in a singleton set. Suppose that at some stage we have partitioned $v_1, \ldots, v_m$ (where m < n) into k sets. If v_{m+1} is adjacent to none of the vertices in some set in the partition, partition $v_1, \ldots, v_{m+1}$ into at most k sets (looking only at the subgraph H of G_n induced by $v_1, \ldots, v_{m+1}$). Otherwise partition $v_1, \ldots, v_{m+1}$ into at most k+1 sets (again looking only at H). The simple sequential colouring algorithm SA and 'SA with interchanges' are examples of sequential partitioning algorithms.

For any sequential partitioning algorithm A, let A(G) denote the number of sets into which A partitions the vertex-set of a graph G, and let A*(G) denote the maximum value of A(H) over all the vertex re-orderings H of G. Note that $(SA)*(G) = \psi_c(G)$ for any graph G. Thus Theorem 5 below implies (6), and thus yields Theorem 3. Also Lemma 6 yields the similar looking result Lemma 4. (Theorem 5 and Lemma 6 do not, however, seem to tell us about the 'canonical pseudo achromatic number'.)

<u>Theorem 5.</u> Let A be any sequential partitioning algorithm, and let $\epsilon > 0$. Then, for almost all graphs G_n,

$$A*(G_n) < (1 + \epsilon) \, SA(G_n). \qquad \square$$

This theorem follows immediately from (4) and Lemma 6 below (which is stated in more detail then is necessary just to deduce the theorem), since there are only n! vertex re-orderings of a graph G_n.

<u>Lemma 6.</u> Let $4 \leq v(n) < \log n / \log\log n$, and let

$$k(n) = \left\lceil \frac{n \log(q^{-1})}{\log n} \left(1 - v(n) \frac{\log\log n}{\log n}\right)^{-1} \right\rceil.$$

Then, for n sufficiently large,

$$P\{A(G_n) \geq k(n)\} < \exp\{-n(\log n)^{v-4}\}.$$

($\lceil x \rceil$ denotes the least integer at least x.)

Proof. Suppose that at some stage the algorithm A has partitioned the $m < n$ vertices $v_1, \ldots, v_m$ into j sets of sizes $k_1, \ldots, k_j$. Then, as noted in the proof of Theorem 4(ii) of [2], the probability that A introduces an extra set when considering the next vertex v_{m+1} is at most

$$\prod_{i=1}^{j} (1 - q^{k_i}) \leq (1 - q^{m/j})^j < \exp\{-jq^{n/j}\}.$$

Hence

$$P\{A(G_n) \geq j+1 | A(G_n) \geq j\} < n \exp\{-jq^{n/j}\}.$$

Note that $f(j) = n \exp\{-jq^{n/j}\}$ is a decreasing function of j.
Now let

$$k'(n) = \frac{n \log(q^{-1})}{\log n} \left[1 - (v-1) \frac{\log\log n}{\log n}\right]^{-1},$$

and note that

$$k(n) - k'(n) > n(\log(q^{-1}))(\log\log n)(\log n)^{-2},$$

and

$$f(k') \leq \exp\{-(\log n)^{v-2}(\log(q^{-1}))(1 + o(1))\}.$$

Then

$$P\{A(G_n) \geq k(n)\}$$

$$\leq \prod_{j=k'}^{k-1} P\{A(G_n) \geq j+1 | A(G_n) \geq j\}$$

$$\leq f(k')^{k-k'-1}$$

$$\leq \exp\{-n(\log n)^{v-4}\} \quad \text{for n sufficiently large.} \quad \square$$

Lemma 6 should be compared with Lemma 4. Both are an improvement on a result of Erdös and Bollobás: in Theorem 4(ii) of [2] they obtained the upper bound

$$\exp\{-(\log n)^{v-2}\},$$

instead of our

$$\exp\{-n(\log n)^{v-4}\}.$$

We have seen that the average behaviour of graph colouring heuristics is much better than the extreme behaviour. Also, while the simple sequential colouring algorithm SA is quite good (by (3)), it is also quite bad (by (6) and Theorem 5). There are many questions still to be answered, even with our simple probability assumptions.

It has been conjectured in [5] (and in [2]) that, for almost all graphs G_n,

$$SA(G_n)/\chi(G_n) \geq 2 - \epsilon,$$

so that we might hope to find fast colouring heuristics with better asymptotic average behaviour than SA. It would be interesting to know if any of the popular heuristics (such as those in [7]) beats SA in this way, and indeed if any polynomial heuristic does (see also [3]).

REFERENCES

1. A. E. Aho, J. E. Hopcraft, and J. D. Ullman, The Design and Analysis of Computer Algorithms, Addison-Wesley, Reading, Mass., 1974.

2. B. Bollobás and P. Erdös, Cliques in random graphs, Math. Proc. Cambridge Phil. Soc. 80 (1976), 419-427.

3. F. D. J. Dunstan, Sequential colourings of graphs, Proc. Fifth British Combinatorial Conf., Utilitas Mathematica, Winnipeg (1976), 151-158.

4. M. R. Garey and D. S. Johnson, The complexity of near-optimal graph colouring, J. ACM 23 (1976), 43-49.

5. G. R. Grimmett and C. J. H. McDiarmid, On colouring random graphs, Math. Proc. Cambridge Phil. Soc. 77 (1975), 313-324.

6. R. P. Gupta, Bounds on the chromatic and achromatic numbers of complementary graphs, Recent Progress in Combinatorics (ed. W. T. Tutte), Academic Press, New York (1969), 229-235.

7. D. S. Johnson, Worst case behaviour of graph colouring algorithms, Proc. Fifth Southeastern Conf. on Combinatorics, Graph Theory and Computing, Utilitas Mathematica, Winnipeg (1974), 513-538.

8. D. W. Matula, G. Marble and J. D. Isaacson, Graph colouring algorithms, Graph Theory and Computing (ed. R. C. Read), Academic Press, New York (1972), 109-122.

9. C. J. H. McDiarmid, Determining the chromatic number of a graph, Stanford University Report STAN-CS-76-576, 1976.

C. McDiarmid
London School of Economics
Houghton Street
London WC2A 2AE

C St J A Nash-Williams
Acyclic detachments of graphs

Loosely speaking, we might define a detachment of a graph G to be a graph F obtained from G by splitting each vertex of G into one or more vertices as illustrated in Figure 1. Thus F has the same edges as G; and an edge joining two vertices ξ, η in G becomes, in F, an edge joining one of the vertices into which ξ has been split to one of the vertices into which η has been split. A formal definition is as follows: F is a <u>detachment</u> of G if $E(F) = E(G)$ and there exists a function p from $V(F)$ onto $V(G)$ such that, for each $\lambda \in E(G)$, the vertices joined by λ in G are the images under p of the vertices joined by λ in F.

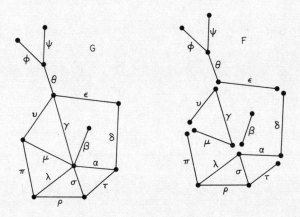

Figure 1

The concept of detachment can be used to restate a number of known results about trails in graphs in a way which may suggest new problems. In the terminology of [2], a <u>trail</u> is a journey in a graph, starting (if at all) at a vertex and terminating (if at all) at a vertex, and not traversing any edge more than once, although it may pass through a vertex more than once. [The words "(if at all)" serve to permit infinite as well as finite trails.] A trail in a graph G is called an Euler trail of G if every vertex and edge of G is in the trail. A set of trails in G will be said to cover G if every vertex and edge of G is in at least one of these trails. Possibly the best known elementary theorem of graph theory states that a non-empty finite graph G has a closed Euler trail (i.e. an Euler trail which starts and terminates at the same vertex) if and only if it is connected and all of its vertices have even valencies. Saying that G has a closed Euler trail is equivalent to saying that some detachment of G is a circuit. For instance, the fact that the graph in Figure 2(i) has a closed Euler trail in which the edges are traversed in the order indicated by the numbering is essentially equivalent to the fact that the detachment of this graph in Figure 2(ii) is a circuit.

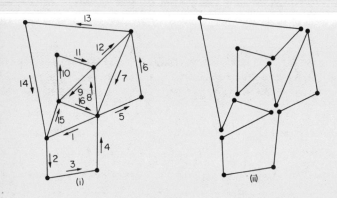

Figure 2

Likewise, asserting that an infinite graph has a one-way infinite Euler trail (i.e. an Euler trail which starts at some vertex but never terminates) is equivalent to asserting that some detachment of the graph is a one-way infinite path, and asserting that an infinite graph has a two-way infinite Euler trail (i.e. an Euler trail going off to infinity at both ends) is equivalent to asserting that it has a detachment which is a two-way infinite path. Thus the theorems of Erdös, Gallai and Vázsonyi [6, Section 3.2] which characterize graphs with one-way or two-way infinite Euler trails are tantamount to characterizing graphs of which some detachment is a one-way or two-way infinite path.

Specified trails in a graph are said to be __edge-disjoint__ if no two of these trails share an edge. In [4], I proved the following (fairly easy) theorem.

__Theorem 1__. Let G be a graph. There exists a set of edge-disjoint two-way infinite trails in G which covers G if and only if G has no vertex of odd finite valency and no finite component. □

This is equivalent to saying that G has a detachment each of whose components is a two-way infinite path if and only if G has no vertex of odd valency and no finite component.

A __forest__ or __acyclic graph__ is a graph in which there are no circuits, i.e. a graph each of whose components is a tree. Let a graph be called __m-regular__ if each of its vertices has valency m. A two-way infinite path is the same thing as a 2-regular tree; and a graph is a 2-regular forest if and only if each of its components is a two-way infinite path. Thus the theorem of Erdös, Gallai and Vázsonyi concerning two-way infinite Euler trails establishes necessary and sufficient conditions for a graph to have a detachment which is a 2-regular tree; and Theorem 1 establishes necessary and sufficient conditions for a graph to have a detachment which is a 2-regular forest. This suggests the more general questions:
(a) which graphs have a detachment which is a m-regular tree?
(b) which graphs have a detachment which is an m-regular forest?
At present, I can only report an answer (Theorem 2 below) to question (b) for the restricted case of locally finite graphs (i.e. graphs in which every vertex has finite valency), and even that answer requires only fairly easy arguments to deduce it from a theorem of J. Edmonds concerning matroids. (At the time of giving the lecture on which this paper is based, I had not noticed that one could use matroids in this problem and I had in mind a purely graph-theoretic proof of Theorem 2.)

__Notation__. If G is a graph and X is a subset of V(G), then E_X will denote the set of those edges of G which join two (not necessarily distinct) elements of X. The valency of a vertex ξ in a graph H will be denoted by $v_H(\xi)$; but the suffix may be dropped when the valency is measured in a graph denoted by the particular symbol G; i.e. $v(\xi)$ always means $v_G(\xi)$.

Theorem 2. Let G be a locally finite graph and m be a positive integer. Then G has an m-regular acyclic detachment if and only if the valencies of the vertices of G are all non-zero multiples of m and

$$m(|E_X| + 1) \le \sum_{\xi \in X} v(\xi)$$

for every non-empty finite subset X of $V(G)$. □

Definitions. Let F, G be graphs. Suppose that $E(F) = E(G)$ and there exists a function p from $V(F)$ onto $V(G)$ such that, for every edge λ of F, the vertices joined by λ in G are the images under p of the vertices joined by λ in F. Then we shall say that p is a <u>coalescence</u> of F onto G and that F is a detachment of G in which each vertex ξ of G is <u>split</u> into the vertices of F which belong to $p^{-1}(\{\xi\})$.

If X is a finite subset of the domain of an integer-valued function f, we will write $f.X$ for $\Sigma f(\xi)$ ($\xi \in X$).

To prove Theorem 2, we will first prove the following theorem.

Theorem 3. Let G be a locally finite graph and b be a function from $V(G)$ into the set of positive integers. There exists an acyclic detachment F of G such that each $\xi \in V(G)$ is split into $b(\xi)$ vertices of F if and only if

$$|E_X| + 1 \le b.X \tag{1}$$

for every non-empty finite subset X of $V(G)$.

Furthermore, if (1) is true and if, for each $\xi \in V(G)$, a sequence s_ξ of $b(\xi)$ positive integers with sum $v(\xi)$ is given, then F can be chosen so that, for each $\xi \in V(G)$, the F-valencies of the vertices into which ξ is split are the terms of s_ξ. □

The proof of Theorem 3 requires a theorem of Edmonds concerning matroids. A <u>matroid</u> is an ordered pair $(M, \underline{I})$ such that M is a finite set and $\underline{I}$ is a set of subsets of M, called <u>independent sets</u>, which satisfies the following conditions:

(i) the empty set is independent;

(ii) every subset of an independent set is independent;

(iii) for each subset A of M, all maximal independent subsets of A have the same cardinality. (This cardinality is called the <u>rank</u> of A.)

For example, if G is a finite graph and $\underline{F}_G$ is the set of all acyclic subgraphs of G and $\underline{I}_G$ denotes $\{E(F): F \in \underline{F}_G\}$ then $(E(G), \underline{I}_G)$ is a matroid, called the <u>matroid of G</u>.

Edmonds has proved the following theorem [1, Section 69],[7, page 130] which relates the maximum of the cardinalities of the common independent sets of two matroids to the rank functions of these matroids.

<u>Theorem 4</u>. Let $(M, \underline{I}_1)$, $(M, \underline{I}_2)$ be two matroids with the same underlying set M. Then

$$\max_{I \in \underline{I}_1 \cap \underline{I}_2} |I| = \min_{A \subseteq M} (r_1(A) + r_2(M \setminus A)),$$

where $r_j(Z)$ denotes the rank in $(M, \underline{I}_j)$ of a subset Z of M. □

<u>Definition</u>. A <u>spanning subgraph</u> of a graph G is a subgraph S of G such that $V(S) = V(G)$; if in addition S is a forest, we may call it a <u>spanning subforest</u> of G.

<u>Proof of Theorem 3</u>. If there exists an acyclic detachment F of G such that each $\xi \in V(G)$ splits into $b(\xi)$ vertices of F, then there is a coalescence p of F onto G such that $|p^{-1}(\{\xi\})| = b(\xi)$ for every $\xi \in V(G)$. If X is a non-empty finite subset of V(G), let F_X be the subgraph of F such that $V(F_X) = p^{-1}(X)$ and $E(F_X) = E_X$. Then F_X is a non-empty finite forest, and therefore $|E(F_X)| + 1 \leq |V(F_X)|$; i.e. $|E_X| + 1 \leq |p^{-1}(X)| = b.X$.

For the converse proof, assume (1). Select disjoint sets $\Omega\xi$, $\xi \in V(G)$ such that $|\Omega\xi| = b(\xi)$ for each $\xi \in V(G)$. Let Γ be a graph such that $V(\Gamma) = \bigcup \Omega\xi$ ($\xi \in V(G)$), $E(\Gamma)$ is the union of disjoint sets $\Omega\lambda$, $\lambda \in E(G)$ and, for each edge λ of G,

(i) if λ joins distinct vertices ξ, η, then $\Omega\lambda$ is the set of edges of a complete bipartite subgraph of Γ whose set of vertices is $\Omega\xi \cup \Omega\eta$, each element of $\Omega\xi$ being adjacent in this subgraph to each element of $\Omega\eta$,

(ii) if λ is a loop incident with a vertex ξ of G, then $\Omega\lambda$ is the set of edges of a complete subgraph of Γ whose set of vertices is $\Omega\xi$.

For any subgraph H of G, let ΩH be the subgraph of Γ such that $V(\Omega H) = \bigcup \Omega\xi$ ($\xi \in V(H)$), $E(\Omega H) = \bigcup \Omega\lambda$ ($\lambda \in E(H)$), and let $\Phi(H)$ be the set of all spanning subforests P of ΩH such that $|E(P) \cap \Omega\lambda| = 1$ for every $\lambda \in E(H)$.

Consider first a finite subgraph H of G. Let $(E(\Omega H), \underline{I}_1)$ be the matroid of ΩH (i.e. let $\underline{I}_1 = \underline{I}_{\Omega H}$), and let $(E(\Omega H), \underline{I}_2)$ be the matroid such that a subset I of $E(\Omega H)$ belongs to $\underline{I}_2$ if and only if $|I \cap \Omega\lambda| \leq 1$ for each edge λ of H. Let $r_j(Z)$ denote the rank in $(E(\Omega H), \underline{I}_j)$ of any subset Z of $E(\Omega H)$. Let A be a subset of $E(\Omega H)$. Let $D = \{\lambda \in E(H): A \cap \Omega\lambda = \emptyset\}$ and let $H_1, \ldots, H_h$ be the components of H-D. (H-D denotes the spanning subgraph of H whose set of edges is $E(H)\setminus D$.) For any non-empty finite connected subgraph K of G, clearly ΩK is connected and so has a spanning tree T such that

$$|E(T)| = |V(\Omega K)| - 1 = b.V(K) - 1 \geq |E_{V(K)}| \geq |E(K)|.$$

Therefore we can find a spanning tree T_i of ΩH_i such that $|E(T_i)| \geq |E(H_i)|$ for $i = 1, \ldots, h$ and so $\Omega(H - D)$ has a spanning subforest $P = T_1 \cup \ldots \cup T_h$ with $|E(P)| \geq \sum|E(H_i)| = |E(H)| - |D|$. Since $E(\Omega(H - D)) \subset E(\Omega H)\setminus A$, we infer that

$$r_1(E(\Omega H)\setminus A) \geq r_1(E(\Omega(H - D))) \geq |E(P)| \geq |E(H)| - |D|.$$

It is clear that $r_2(A) = |D|$ since a maximal subset of A belonging to $\underline{I}_2$ will include just one representative of $A \cap \Omega\lambda$ for each $\lambda \in D$. Consequently $r_1(E(\Omega H)\setminus A) + r_2(A) \geq |E(H)|$. This inequality has been proved for all subsets A of $E(\Omega H)$: therefore, by Theorem 4, there exists $I \in \underline{I}_1 \cap \underline{I}_2$ such that $|I| \geq |E(H)|$. Let Q be the spanning subgraph of ΩH such that $E(Q) = I$. Since $|I| \geq |E(H)|$, and $I \in \underline{I}_2$, it follows that $|I \cap \Omega\lambda|$ must be exactly 1 for each $\lambda \in E(H)$. Since $I \in \underline{I}_1$, it follows that Q is a forest. Hence $Q \in \Phi(H)$. We have thus proved that $\Phi(H) \neq \emptyset$ for each finite subgraph H of G.

Now let H be a denumerably infinite subgraph of G. [By calling H <u>denumerably infinite</u> we mean that the set $V(H) \cup E(H)$ is denumerably infinite; but in fact this is equivalent to V(H) being denumerably infinite because H is a subgraph of the locally finite graph G.] We can express H as the union of an infinite sequence H_1, H_2, ... of finite subgraphs such that $H_1 \subset H_2 \subset \ldots$. For each positive integer n, $\Phi(H_n)$ is non-empty by the argument of the preceding paragraph. Therefore the set $\Phi(H_1) \cup \Phi(H_2) \cup \Phi(H_3) \cup \ldots$ is infinite; but clearly each of its elements contains an element of $\Phi(H_1)$, which is finite. Therefore $\Phi(H_1)$ must have an element Q_1 which is contained in infinitely many elements of $\Phi(H_2) \cup \Phi(H_3) \cup \Phi(H_4) \cup \ldots$. Each of these elements of $\Phi(H_2) \cup \Phi(H_3) \cup \Phi(H_4) \cup \ldots$ must contain an element of $\Phi(H_2)$ which contains Q_1. Since $\Phi(H_2)$ is finite, it must have an element Q_2 which contains Q_1 and is contained in infinitely many elements of $\Phi(H_3) \cup \Phi(H_4) \cup \Phi(H_5) \cup \ldots$. Each of these elements of $\Phi(H_3) \cup \Phi(H_4) \cup \Phi(H_5) \cup \ldots$ must contain an element of $\Phi(H_3)$ which contains Q_2. Since $\Phi(H_3)$ is finite, it must have an element Q_3 which contains Q_2 and is contained in infinitely many elements of $\Phi(H_4) \cup \Phi(H_5) \cup \Phi(H_6) \cup \ldots$. Continuing this argument, we can produce an infinite sequence $Q_1 \subset Q_2 \subset Q_3 \subset \ldots$ such that $Q_n \in \Phi(H_n)$ for each n. Clearly $Q_1 \cup Q_2 \cup \ldots \in \Phi(H)$, and so $\Phi(H) \neq \emptyset$.

We have now proved that $\Phi(H) \neq \emptyset$ for each finite or denumerably infinite subgraph H of G. In particular, $\Phi(C) \neq \emptyset$ for each component C of G, since components of a locally finite graph are finite or denumerably infinite. If we select a $Q_C \in \Phi(C)$ for each component C of G, then the union of all these forests Q_C will be a spanning subforest F* of Γ such that $|E(F^*) \cap \Omega\lambda| = 1$ for each edge λ of G. By replacing the element of $E(F^*) \cap \Omega\lambda$ by λ for each $\lambda \in E(G)$, we can convert F* into an acyclic detachment F of G such that each $\xi \in V(G)$ is split in F into the $b(\xi)$ elements of $\Omega\xi$, thus completing the proof of the first part of Theorem 3.

Now suppose that for each $\xi \in V(G)$ we are given a sequence s_ξ of $b(\xi)$ positive integers with sum $v(\xi)$. Since $|\Omega\xi| = b(\xi)$, we can construct a function u from $V(\Gamma)$ into the set of positive integers such that, for each $\xi \in V(G)$, the numbers $u(\alpha)$, $\alpha \in \Omega\xi$ form the sequence s_ξ when arranged in a suitable order.

For any component C of G let $\Phi_u(C)$ denote the set of all elements Q of $\Phi(C)$ such that $v_Q(\alpha) = u(\alpha)$ for every $\alpha \in V(\Omega C)$. If $Q \in \Phi(C)$ and $\xi \in V(C)$, define a $\underline{\xi\text{-modification}}$ of Q to be an $R \in \Phi(C)$ such that
(i) $E(R) \cap \Omega\lambda = E(Q) \cap \Omega\lambda$ for every edge λ of G which is not incident with ξ,
(ii) for every pair λ, η such that λ is an edge of G joining ξ to a vertex

$\eta \neq \xi$, the element of $E(R) \cap \Omega\lambda$ and the element of $E(Q) \cap \Omega\lambda$ are incident with the same element of $\Omega\eta$.

(Thus vertices in $V(C)\setminus\Omega\xi$ have the same valencies in R as in Q.) If R is a ξ-modification of Q and $v_R(\alpha) = u(\alpha)$ for each $\alpha \in \Omega\xi$, we shall say that R is a <u>ξu-modification</u> of Q.

We prove first that if ξ is a vertex of a component C of G and $Q \in \Phi(C)$ then Q has a ξu-modification. To see this, observe first that Q has at least one ξ-modification, viz. Q itself, and therefore we can choose a ξ-modification R of Q for which $\sum |v_R(\alpha) - u(\alpha)|$ ($\alpha \in \Omega\xi$) assumes the least value. Now $v(\xi)$ is equal to the sum of the terms of s_ξ which is equal to $u.\Omega\xi$, and $|E(R) \cap \Omega\lambda| = 1$ for each $\lambda \in E(C)$. Hence, if L is the set of loops joining ξ to itself in G and W is the set of all other edges incident with ξ in G, then

$$v_R . \Omega\xi = 2 \sum_{\lambda \in L} |E(R) \cap \Omega\lambda| + \sum_{\lambda \in W} |E(R) \cap \Omega\lambda|$$
$$= 2|L| + |W| = v(\xi) = u.\Omega\xi.$$

Therefore either R is a ξu-modification of Q or there exist $\theta_1, \theta_2 \in \Omega\xi$ such that

$$v_R(\theta_1) > u(\theta_1), \quad v_R(\theta_2) < u(\theta_2); \tag{2}$$

but we will show that the latter alternative leads to a contradication. Assume that θ_1, θ_2 exist. Since R is a forest, there is at most one $\theta_1\theta_2$-path in R. Since $v_R(\theta_1) > u(\theta_1) \geq 1$, we can select an edge ε of R incident with θ_1 such that, if there is a $\theta_1\theta_2$-path in R, then ε is not in this path. Let λ be the edge of G such that $\varepsilon \in \Omega\lambda$ and let λ join ξ to the vertex η of G. Then ε joins θ_1 to some $\phi \in \Omega\eta$. By the definition of Γ, there is an $\varepsilon' \in \Omega\lambda$ which joins ϕ to θ_2. Let S be the spanning subgraph of ΩC obtained from R by removing ε and adding ε'. It is easily seen that S is a ξ-modification of Q and

$$v_S(\theta_1) = v_R(\theta_1) - 1, \; v_S(\theta_2) = v_R(\theta_2) + 1,$$
$$v_S(\alpha) = v_R(\alpha) \; (\alpha \in \Omega\xi\setminus\{\theta_1, \theta_2\}), \tag{3}$$

and it follows from (2) and (3) that

$$\sum_{\alpha \in \Omega\xi} |v_S(\alpha) - u(\alpha)| < \sum_{\alpha \in \Omega\xi} |v_R(\alpha) - u(\alpha)|,$$

contradicting the choice of R. This contradiction shows that R is a ξu-modification of Q, as required.

Consider any component C of G. Since we have seen that $\Phi(C) \neq \emptyset$, we can select a $Q_0 \in \Phi(C)$. Since G is locally finite, V(C) is finite or denumerably infinite. If V(C) is finite, let $V(C) = \{\xi_1, \ldots, \xi_n\}$. By the argument of the preceding paragraph, we can find a $\xi_1 u$-modification Q_1 of Q_0 and then a $\xi_2 u$-modification Q_2 of Q_1 and so on, ending with a $\xi_n u$-modification Q_n of Q_{n-1}. Clearly $Q_n \in \Phi_u(C)$ and so $\Phi_u(C) \neq \emptyset$. If V(C) is denumerably infinite, let $V(C) = \{\xi_1, \xi_2, \ldots\}$. By the argument of the preceding paragraph, we can find a $\xi_1 u$-modification Q_1 of Q_0 and then a $\xi_2 u$-modification Q_2 of Q_1 and then a $\xi_3 u$-modification Q_3 of Q_2 and so on ad infinitum. Let Q_ω be the spanning subgraph of ΩC such that an edge of ΩC belongs to $E(Q_\omega)$ if and only if it belongs to $E(Q_h)$ for all sufficiently large integers h. If $\lambda \in E(C)$ then λ must join ξ_i to ξ_j for some i, j, and so $E(Q_h) \cap \Omega\lambda$ will be the same set of cardinality 1 for all integers $h \geq \max(i,j)$. Therefore $|E(Q_\omega) \cap \Omega C| = 1$ for each edge λ of C. If Q_ω contained a circuit Z then each edge θ of Z would be in Q_i for all integers i greater than or equal to some integer k_θ, and so Z would be a circuit in Q_i for all integers $i \geq \max k_\theta$, $\theta \in E(Z)$, which is impossible since Q_i is a forest for each non-negative integer i. Therefore Q_ω contains no circuits and is a forest. Clearly each vertex α of ΩC has valency $u(\alpha)$ in Q_ω. Hence $Q_\omega \in \Phi_u(C)$ and so $\Phi_u(C) \neq \emptyset$.

We have now proved that $\Phi_u(C)$ is non-empty for each finite or infinite component C of G. If we select $Q_C \in \Phi_u(C)$ for each component C of G, then the union of all these forests Q_C will be a spanning subforest F* of Γ such that $|E(F*) \cap \Omega\lambda| = 1$ for each edge λ of G and $v_{F*}(\alpha) = u(\alpha)$ for each $\alpha \in V(\Gamma)$. If we convert F* into a detachment F of G as before, then F will be an acyclic detachment of G in which each $\xi \in V(G)$ is split into the elements of $\Omega\xi$, whose F-valencies are the same as their F*-valencies, which are the numbers $u(\alpha), \alpha \in \Omega\xi$, i.e. the terms of s_ξ. This completes the proof of Theorem 3. □

Proof of Theorem 2. If G has an m-regular acyclic detachment F, then any vertex ξ of G splits into some positive number b_ξ of vertices of F and so $v(\xi)$ must be the non-zero multiple $b_\xi m$ of m. Moreover, since F is a detachment of G in which each vertex ξ of G is split into $v(\xi)/m$ vertices, it follows from Theorem 3 that

$$|E_X| + 1 \leq \sum_{\xi \in X} \frac{v(\xi)}{m}$$

for each non-empty finite subset X of V(G).

For the converse argument, assume that the valencies of the vertices of G are non-zero multiples of m and that $m(|E_X| + 1) \leq \sum_{\xi \in X} v(\xi)$ for every non-empty finite subset X of V(G). Apply Theorem 3, taking s_ξ to be the sequence m, m, m, ..., m with $v(\xi)/m$ terms for each $\xi \in V(G)$, and we find that G has an m-regular acyclic detachment. □

Additional complications clearly arise if we try to extend this work to graphs which are not locally finite. For instance, if a graph G has a vertex ξ_r and edges λ_r, μ_r for every integer r, another vertex η, and no further vertices or edges, and if λ_r joins ξ_r to ξ_{r+1} and μ_r joins ξ_r to η for every integer r, then G has no 3-regular acyclic detachment although it satisfies (for m = 3) the conditions of Theorem 2 other than local finiteness. A study of examples also indicates that complications arise when we seek to characterise graphs of which some detachment is an m-regular tree. I believe that progress could be made on these problems; but it might be of somewhat reduced interest if the statements and proofs of the results have to be too cumbersome. Another result proved in [3] (and stated without proof in [5]) gives the necessary and sufficient conditions on a graph G and number n for the existence in G of n edge-disjoint two-way infinite trails which together cover G. The proof in [3] was lengthy, and a reasonable challenge might be to ask whether a better proof can be constructed on the lines of the argument in this paper.

REFERENCES

1. J. Edmonds, Submodular functions, matroids, and certain polyhedra, Combinatorial Structures and their Applications (ed. R. K. Guy, H. Hanani, N. Sauer and J. Schönheim), Gordon and Breach, New York (1970), 69-87; MR 42-5828.

2. F. Harary, Graph Theory, Addison-Wesley, Reading, Mass., 1969; MR 41-1566.

3. C.St.J. A. Nash-Williams, Decomposition of graphs into infinite chains, Ph.D. thesis, Cambridge, 1958.

4. C.St.J. A. Nash-Williams, Decomposition of graphs into closed and endless chains, Proc. London Math. Soc. (3) 10 (1960), 221-238; MR 22-4642.

5. C.St.J. A. Nash-Williams, Decomposition of graphs into two-way infinite paths, Canad. J. Math. 15 (1963), 479-485; MR 27-741.

6. O. Ore, Theory of Graphs, American Mathematical Society, Providence, R.I., 1962; MR 27-740.

7. D. J. A. Welsh, Matroid Theory, Academic Press, London 1976.

C.St.J. A. Nash-Williams
Department of Mathematics
The University of Reading
Reading
Berks RG6 2AX
England

F Piper
Unitary block designs

1. INTRODUCTION

A <u>unitary block design</u> (or <u>unital</u>) with parameter q is a $2 - (q^3 + 1, q + 1, 1)$ design. Thus it is an incidence structure of points and lines such that
(i) there are $q^3 + 1$ points;
(ii) each line contains $q + 1$ points;
(iii) any pair of distinct points on a unique common line.
It follows immediately from the definition that a unital with parameter q has $q^2(q^2 - q + 1)$ lines, and that each point lies on q^2 lines.

For any prime power q, the absolute points and non-absolute lines of a unitary polarity of $PG(2, q^2)$ form a unital. (Hence the name "unital", and also the initial motivation and interest for studying these designs.) We call such a unital <u>classical</u>. Since all unitary polarities of $PG(2, q^2)$ are conjugate in $P\Gamma U(3, q^2)$, all classical unitals with the same parameter are isomorphic; so we will talk of <u>the</u> classical unital with parameter q, and denote it by $C(q)$. Any automorphism of $PG(2, q^2)$ which commutes with a unitary polarity acts as an automorphism of the corresponding $C(q)$. Thus $P\Gamma U(3, q^2)$ acts as an automorphism group of $C(q)$. Furthermore, it is straightforward to verify that $P\Gamma U(3, q^2)$ acts doubly transitively on the points of $C(q)$.

There is a second known class of unitals which have doubly transitive automorphism groups. For any $m > 0$, if $q = 3^{2m+1}$, then Lüneburg [7] showed that there is a unital with parameter q which has the Ree group as a doubly transitive automorphism group. We call such a unital a <u>Ree unital</u>, and denote it by $R(q)$. It is easily verified that $R(q) \neq C(q)$.

In this paper we will discuss the historical development of the study of unitals, with particular reference to the discovery of new families. We also discuss the relation between projective planes and unitals. Of course, one of the major unanswered questions is: for which values of q are there unitals with parameter q? So far all known unitals have prime power q (as do planes). None of the constructions listed here can possibly give a

unital with a new parameter because they all begin with a projective plane
of order q^2, determine a unital with parameter q in the plane, and then
define other unitals with the same point set.

2. UNITALS IN FINITE PROJECTIVE PLANES

In 1946 Baer [1] established a lower bound of n + 1 for the number of
absolute points of a polarity of a projective plane of order n. He also
showed that, when this bound is attained, the absolute points form an oval
if n is odd, and are collinear if n is even. In 1970 Seib [10]
established an upper bound and showed:

Theorem 1. If θ is a polarity of a finite projective plane of order n
having $a(\theta)$ absolute points, then $n + 1 \leq a(\theta) \leq n^{3/2} + 1$. □

Theorem 2. If θ is a polarity of a finite projective plane of order m^2
having $m^3 + 1$ absolute points, then the absolute points and non-absolute
lines of θ form a unital with parameter m. □

In view of Theorem 2, we call a polarity θ of a plane of order n *unitary*
if $a(\theta) = n^{3/2} + 1$.

When one is studying polarities of finite projective planes, it is
usually easy to count the number of absolute points of a given polarity.
However, determining the configuration which they form is often very
difficult. Thus Theorem 2 is very powerful. It means that any polarity
with the correct number of absolute points must give rise to a unital.

In his thesis, Ganley carried out a systematic study of polarities
of finite projective planes and discovered many examples of unitary
polarities. Most of the work was concerned with translation planes -
that is, planes coordinatized by quasifields.

If Q is any quasifield, then Q determines an affine (translation) plane
A(Q) as follows: The points of A(Q) are the ordered pairs (x,y) with x,y
in Q. The lines of A(Q) are the ordered pairs [m,k] with m,k in Q, and
each element k of Q defines a further line [k]. Incidence is given by:
(x,y) is on [m,k] if y = mx + k, and (x,y) is on [k] if x = k. The affine
plane A(Q) then uniquely determines a projective plane P(Q). (For details,
see Hughes and Piper [6].) Elementary considerations show that P(Q) cannot
admit a polarity unless Q is a semifield. Ganley [4] proved:

Theorem 3. If Q is a finite commutative semifield admitting a non-trivial involutory automorphism, then P(Q) admits a unitary polarity. □

One special case of a finite commutative semifield is a field, and it is well known that GF(q) admits a non-trivial involutory automorphism if and only if q is a square. Thus a well-known corollary of Theorem 3 is the existence of unitals in PG(2, q^2). However there are many finite commutative semifields which are not fields, and thus Theorem 3 establishes the existence of unitals in non-desarguesian projective planes.

The way in which Ganley discovered these unitals was one natural generalisation of the way in which classical unitals arise - that is, as the absolute points of polarities of finite projective planes. More recently, Buekenhout [2] has found another natural generalisation.

If Q is a quasifield of order q^2 whose kernel contains GF(q), then the points of A(Q) may be regarded as the vectors of $V_4(q)$ and the lines are then certain 2-dimensional subspaces. Thus one may regard all translation planes of order q^2 and kernel containing GF(q) as having the same set of points, and, by regarding a unital in PG(2, q^2) as the set of points satisfying a Hermitian form and then looking at the partial quadric with the same equation in $V_4(q)$, Buekenhout showed that, under certain circumstances, the points which form a unital in PG(2, q^2) form a unital in the other translation planes.

Theorem 4. If Q is a finite quasifield of q^2 elements whose kernel contains GF(q) then P(Q) contains a unital with parameter q. □

There are many such quasifields and, more interestingly, many which are not semifields. If Q is not a semifield, then P(Q) cannot admit a polarity. Thus Buekenhout has established the existence of unitals in planes which do not have polarities.

We conclude this section by observing that, if $q \neq 2$, PG(2, q^2) contains a unital which does not arise as the set of absolute points of a polarity. This was shown by Metz [8], although it had previously been well known for even q.

3. THE ISOMORPHISM PROBLEM

In Section 2 we were essentially discussing a pair of structures (P,U) where P is a projective plane of order q^2 containing a unital U with parameter q. We showed many instances where the planes are non-isomorphic, but so far we have said nothing about the possible isomorphisms between the unitals. Very little is known about the general problem of deciding whether two unitals are isomorphic although, as we shall see, it is usually possible to distinguish C(q) from the other unitals with parameter q. However, before we start discussing the isomorphism problem, it is worth mentioning some problems about the relation between unitals and planes.

<u>Problem 1</u>. If U is a unital with parameter q, must there be a projective plane of order q^2 containing U? (The answer to this is almost certainly 'no'. Lüneburg [7] showed that R(q) cannot be embedded in a plane P of order q^2 in such a way that the Ree group acts on P. More recently, Dorber [3] has constructed a number of unitals with parameter 3, and it seems extremely unlikely that they can all be embedded in planes of order 9.)

<u>Problem 2</u>. If a unital with parameter q can be embedded in a plane of order q^2, is the plane (unital) uniquely determined by the unital (plane)? (Answering this would certainly help with determining possible isomorphisms between the unitals in §2.)

<u>Problem 3</u>. If U is a unital with parameter q in a plane P of order q^2, then comp(U), the complement of U in P, is the dual of a unital. If we write $\bar{U}$ for the dual of comp(U), then when is $U \simeq \bar{U}$? (Is it true, for instance, that $U \simeq \bar{U}$ if and only if U is the set of absolute points of a polarity of P?)

The study of unitals as abstract designs was really begun by group theorists. O'Nan [9] showed that the full automorphism group of C(q) is PΓU(3, q^2). As a result of his work we know that if (P,U) is a pair with PΓU(3, q^2) ⊆ Aut P acting on U, then P ≃ PG(2, q^2) and U ≃ C(q) (see Taylor [11]), and that if U is a unital with Aut U ≃ PΓU(3, q^2) then U ≃ C(q).

However, if we are given a pair (P,U), although it is often easy to determine the subgroup of Aut P which preserves U, there is no reason for assuming that Aut U is not considerably larger. In fact it is, in general, extremely hard to determine Aut U. Nevertheless, if the unital is not symmetric (in the sense that the configurational structure at every point is not the same), then it cannot admit a transitive automorphism group, and cannot be classical or Ree.

O'Nan noticed that the classical unital does not contain the following configuration, which we will call the O'Nan configuration:

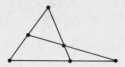

(Note that the O'Nan configuration is merely the dual of the complete graph on 4 points.)

This observation of O'Nan has played a central role in deciding whether two unitals are isomorphic. Ganley [5] showed that his unitals with parameter 3^2 or 5^2 contained O'Nan configurations, and hence were not classical. By using a construction which will be described in §4, Dorber has shown there are at least thirteen non-isomorphic unitals with parameter 3. He established the non-isomorphisms by actually computing the number of O'Nan configurations involving each point in each unital. Of course this approach is not feasible for larger unitals, but it does suggest that the O'Nan configuration will be important in future developments. It is widely conjectured, for instance, that a unital with no O'Nan configurations must be classical.

4. NEW UNITALS FROM OLD: DORBER'S CONSTRUCTION

As we pointed out earlier, it is still not completely clear whether unitals should be studied in conjunction with projective planes or on their own. However the indications are that either there are many unitals which cannot be embedded in planes, or it will be possible to use unitals to construct new planes.

In his thesis Dorber [3] has studied unitals in their own right - that is, without reference to planes. By analysing how the Buekenhout unitals may be obtained from the classical ones, he arrived at the following:

Let U be any unital with parameter q, let ℓ be a line of U, and let $A_1, \ldots, A_{q+1}$ be the points on ℓ.

Definition. U is <u>derivable</u> with respect to ℓ if the lines of U\\ℓ which intersect ℓ can be partitioned into classes $\underline{B}_{ij}$ ($1 \le i, j \le q + 1$) such that
(1) $|B_{ij}| = q - 1$ for all i and j;
(2) if m is a line of U which intersects ℓ, then, for any fixed i,
$$A_i \in m \Longleftrightarrow m \in \underline{B}_{ij} \text{ for some } j;$$
(3) given $P \notin \ell$, and given any fixed j, there exists a unique i such that P is on a line of $\underline{B}_{ij}$.

If U is derivable with respect to the line ℓ, then we can define another unital U* (called the <u>derived unital</u> of U), as follows: the points of U* are the points of U which are not on ℓ, plus q + 1 new points $B_1, \ldots, B_{q+1}$; the lines of U* are the lines of U other than ℓ, plus one new line ℓ*; and incidence in U* is given by

$$A \in m \text{ in } U^* \Longleftrightarrow A \in m \text{ in } U$$

$$B_j \in m \text{ in } U^* \Longleftrightarrow m \in \underline{B}_{ij} \text{ for some i in u}$$

$$B_i \text{ is on } \ell^* \text{ for all i.}$$

It is easy to verify that U* is indeed a unital.

Every classical unital is derivable with respect to any line, and furthermore the derived classical unital is derivable with respect to certain lines other than ℓ*. (Note that if U is derivable with respect to ℓ, then U* is automatically derivable with respect to ℓ*.) Of course, it is difficult to decide if an arbitrary unital is derivable, but often, if a unital is embedded in the plane, it is possible to use the geometry of the plane to establish the derivability of the unital and then to isolate it from the plane and derive it. Repeated use of derivation was the basic tool which enabled Dorber to construct his unitals with q = 3.

If a unital U is derivable with respect to a line ℓ, then, using the notation of the earlier part of this section, we can define a structure S as follows.   There are two types of points and lines:

points of type 1 = points of U not on ℓ

points of type 2 = the classes $\underline{B}_{ij}$ ($1 \leq i \leq q + 1$, $1 \leq j \leq q + 1$)

lines of type 1 = the lines of $U \setminus \ell$ which meet ℓ

lines of type 2 = $A_1, \ldots, A_{q+1}, B_1, \ldots, B_{q+1}$.

Incidence is given by:

a point P of type 1 is on a line m of type 1 in S $\Longleftrightarrow$ P is on m in U.

a point P of type 1 is never incident with a line of type 2 in S.

a point $\underline{B}_{ij}$ of type 2 is on a line m of type 1 in S $\Longleftrightarrow$ m $\in \underline{B}_{ij}$ in U.

a point $\underline{B}_{ij}$ of type 2 is on a line A_h or B_k in S $\Longleftrightarrow$ h = i or k = j.

For any derivable unital U the structure S has $q^3 + q^2 + q + 1$ points and lines. (In fact it is a symmetric $1 - (q^3 + q^2 + q + 1, q + 1, q + 1)$ design.) If however, ℓ does not contain two points of an O'Nan configuration, then something interesting occurs:

<u>Theorem 5</u>. If the unital is derivable with respect to a line ℓ, and if ℓ does not contain two points of an O'Nan configuration in U, then S is a generalised quadrangle. □

Clearly U(q) satisfies the conditions of Theorem 5 with any line chosen as ℓ; and in this case, the generalised quadrangle is in fact regular. It is easy to write down a number of (very unsatisfactory) conditions which the unital must satisfy if the quadrangle is to be regular. When this is the case then S consists of the points and lines of a quadric in PG(4, q). As soon as this occurs then, of course, classical geometry can be used to establish properties of unitals. The type of thing which one hopes might occur is that the existence of no O'Nan configurations will imply that the unital is derivable, and that the generalised quadrangle is regular. This would at least imply that the parameter is a prime power, and might lead to establishing that the unital must be classical.

REFERENCES

1. R. Baer, Polarities in finite projective planes, Bull. Amer. Math. Soc. 52 (1946), 77-93; MR 7-387.

2. F. Buekenhout, Existence of unitals in finite translation planes of order q^2 with a kernel of order q, Geometriae Dedicata 5 (1976), 189-194.

3. G. Dorber, Ph.D. thesis, University of London, to appear.

4. M. J. Ganley, Polarities in translation planes, Geometriae Dedicata 1 (1972), 28-40; MR 46-6157.

5. M. J. Ganley, A class of unitary block designs, Math. Z. 128 (1972), 34-42; MR 47-5718.

6. D. R. Hughes and F. C. Piper, Projective planes, Springer-Verlag, New York, 1973; MR 48-12278.

7. H. Lüneburg, Some remarks concerning the Ree groups of type (G_2), J. Algebra 3 (1966), 256-259; MR 33-1357.

8. R. Metz, On a class of unitals, to appear.

9. M. E. O'Nan, Automorphisms of unitary block designs, J. Algebra 20 (1972), 495-511; MR 45-4995.

10. M. Seib, Unitäre Polaritäten endlicher projectiver Ebenen, Arch. Math. (Basel) 21 (1970), 103-112; MR 42-6710.

11. D. E. Taylor, Unitary block designs, J. Combinatorial Theory (A) 16 (1974), 51-56; MR 48-12270.

Fred Piper
Department of Mathematics
Westfield College
(University of London)
London NW3 7ST

A T White
Strongly symmetric maps

1. INTRODUCTION

Maps, or 2-cell imbeddings of (necessarily) connected graphs into orientable surfaces, have been studied for nearly a century now, with the initial interest deriving from P. J. Heawood's map-colouring conjecture. The Ringel-Youngs proof of the Heawood conjecture [28] involved the determination of the genus parameter for complete graphs, stimulating a flurry of calculations of values of this parameter for other families of graphs. The maximum genus parameter gives the upper bound for the imbedding range of a given graph (the lower bound being given by the genus), and recent work by Xuong [39] has characterized this parameter. Among maps in the interior of the imbedding range, those which have attracted the most interest are the self-dual maps and the symmetrical maps. In this paper we first survey briefly the relevant literature, and then combine the two properties of self-duality and symmetricality (and others as well) to form the class of 'strongly symmetric' maps; we then study this class in detail.

2. DEFINITIONS AND BACKGROUND

We regard a graph G as a finite 1-complex, with vertex- and edge-sets $V(G)$ and $E(G)$ being represented by points and straight-line segments, respectively, in R^3. If there is a group Γ and a generating set $\Delta = \Delta^{-1}$ for Γ (Δ not containing the identity for Γ) such that $V(G) = \Gamma$ and $E(G) = \{[g,h]: g^{-1}h \in \Delta\}$, then G is a Cayley graph, and we write $G = G_\Delta(\Gamma)$. The graph automorphism group Aut G is the set of all permutations $\alpha: V(G) \to V(G)$ preserving edges (that is, if $[v_1,v_2] \in E(G)$, then $[\alpha(v_1),\alpha(v_2)] \in E(G)$), under composition. An anti-automorphism for G is a permutation $\beta: V(G) \to V(G)$ exchanging edges and 'non-edges'; that is, $[v_1,v_2] \in E(G)$ if and only if $[\beta(v_1),\beta(v_2)] \notin E(G)$; we denote the set of all such anti-automorphisms by $\overline{\text{Aut}}\, G$. If a graph G is isomorphic to its complement $\bar{G}$, we say that G is self-complementary. It is clear that G is self-complementary if and only if $\overline{\text{Aut}}\, G \neq \phi$, and that in this case Aut $G \cup \overline{\text{Aut}}\, G$ is a subgroup containing Aut G as a (necessarily normal) subgroup of index two. Self-complementary graphs have been studied in some detail, as the following results indicate. (See also Sachs [29] and Gibbs [14].)

<u>Theorem 2.1</u>   (Ringel [26]). There exists a self-complementary graph of order n if and only if $n \equiv 0$ or $1 \pmod{4}$. □

<u>Theorem 2.2</u>   (Ringel [26]). If $n \equiv 1 \pmod{4}$ and β is an anti-automorphism for a self-complementary graph of order n, then β has exactly one fixed point and every other orbit for the action of β on $V(G)$ has length a multiple of four. □

<u>Theorem 2.3</u>   (Read [25]). There are 36 self-complementary graphs of order 9, 5,600 of order 13, and 11,220,000 of order 17. □

<u>Theorem 2.4</u>   (Rao [24]). If G is a self-complementary graph of order $n \geq 8$ and having minimum degree $\delta \geq \frac{1}{4}n$, then G has a 2-factor (a spanning 2-regular subgraph). □

A graph G is a <u>graphical regular representation</u> of a group Γ if (Aut G, V(G)) is a regular permutation group (that is, transitive with $|\text{Aut } G| = |V(G)|$) and Aut $G \simeq \Gamma$.

<u>Theorem 2.5</u>   (Lim [19]). If G is a graphical regular representation of Γ, then G is not self-complementary. □

We define a <u>map</u> to be a pair (G,ρ), where G is a graph and $\rho = \{\rho_v\}_{v \in V(G)}$ is a collection of cyclic permutations where, for each $v \in V(G)$, ρ_v acts on the neighbourhood $N(v) = \{u : [v,u] \in E(G)\}$.

<u>Example 2.6</u>   Let $G = K_4$, $V(G) = \{1,2,3,4\}$, and $\rho = \{\rho_1, \rho_2, \rho_3, \rho_4\}$, with $\rho_1 = (2,4,3)$, $\rho_2 = (3,4,1)$, $\rho_3 = (1,4,2)$, and $\rho_4 = (1,2,3)$; then (G,ρ) is a map.

If G is a Cayley graph $G_\Delta(\Gamma)$, and if $r : \Delta \to \Delta$ is a cyclic permutation, then we define the <u>Cayley</u> map $M(\Gamma, \Delta, r)$ to be the map (G,ρ), where for each $g \in V(G) = \Gamma$ and $h \in N(g)$,

$$\rho_g(h) = gr(g^{-1}h).$$

Example 2.7. Let $G = K_7$, $V(G) = \{0,1,2,3,4,5,6\}$, and $\rho_i = (1+i, 3+i, 2+i, 6+i, 4+i, 5+i)$; then (G,ρ) is a Cayley map $M(Z_7, Z_7-\{0\}, r)$, where $r = (1,3,2,6,4,5)$.

By a <u>surface</u> S we mean a closed orientable 2-manifold; if S has genus k ($k \geq 0$), we write $S = S_k$. If finitely many identifications, of finitely many points each, are made on a surface, the resulting topological space is called a <u>pseudosurface</u>. A graph G is said to be <u>imbeddable</u> in a surface S_k if the geometric realization in R^3 of G as a finite 1-complex is homeomorphic to a subspace of S_k, the homeomorphism given by (say) $i : G \to S_k$. The components of $S_k - i(G)$ are called <u>regions</u>, and if each region is a 2-cell (that is, homeomorphic to R^2), then i is called a <u>2-cell imbedding</u>. It is well known (see, for example, Heffter [17], Edmonds [12], or Youngs [40]) that maps (G,ρ) coincide with 2-cell imbeddings of G into surfaces S_k of fixed orientation. (For the ρ_v not necessarily cyclic, the correspondence is with pseudosurface imbeddings for which each singular point is the image of a vertex.) For instance, the maps of Examples 2.6 and 2.7 are depicted in Figure 1 (where $k = 0$) and Figure 2 ($k = 1$) respectively. In Figure 2 we also show that the map of K_7 in S_1 is a covering space of a simpler toroidal imbedding (see Gross' theory of voltage graphs [15]).

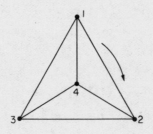

Figure 1

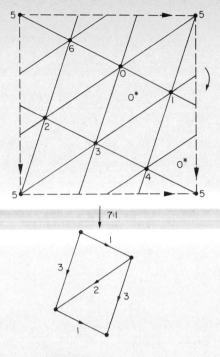

Figure 2

If $i : G \to S_k$ is an imbedding corresponding to a map (G,ρ), then k is said to be the <u>genus of the map</u>. The <u>genus of a graph</u> G, denoted by $\gamma(G)$, is the minimum k such that there exists an imbedding $i : G \to S_k$; the <u>maximum genus</u> of a (connected) graph G, $\gamma_M(G)$, is the maximum k such that there exists a 2-cell imbedding (that is, a map) $i : G \to S_k$. We remark that, as is well known (see Youngs [40], for example), genus imbeddings for connected graphs are also maps; we further observe that every self-complementary graph is connected. Thus very little is sacrificed in general (and nothing is lost in this paper) by restricting our attention to maps.

The value of the genus parameter has been determined for several families of graphs, notably for $G = K_n$, $K_{m,n}$, $K_{n,n,n}$, $K_{n,n,n,n}$, and the n-cubes Q_n. We give only the Complete Graph Theorem here; for a discussion of work on other families, see White and Beineke [37].

<u>Theorem 2.8</u>   (Ringel and Youngs [28]).   The genus of the complete graph K_n of order n is given by

$$\gamma(K_n) = \{\frac{(n-3)(n-4)}{12}\}, \quad n \geq 3. \quad \square$$

The proof uses the generalized Euler equation $v - e + f = 2 - 2k$ to obtain the formula as a lower bound, and then appropriate current graph (see Ringel [27]) or voltage graph (see Gross [15] or White [35]) constructions to attain this bound. The chief importance of this computation is that it establishes the Heawood map colouring theorem (see, for example, Ringel and Youngs [28], or White [36]):

<u>Theorem 2.9.</u> The minimum number of colours sufficing to colour the regions (countries) of every map on S_k is given by

$$\chi(S_k) = [\tfrac{1}{2}(7 + \sqrt{1 + 48k})], \quad k > 0. \quad \square$$

Recent work by Appel and Haken [1], together with Figure 1, shows that Theorem 2.9 holds for $k = 0$ also. Figure 2 illustrates Theorem 2.8 for the case $n = 7$, leading to the result of Theorem 2.9 (in dual form - see below) for the case $k = 1$.

The maximum genus parameter has been characterized recently. Let $\xi_0(H)$ be the number of components of the graph H having an odd number of edges, and for G a connected graph let $\xi(G)$ be the minimum value of $\xi_0(G - E(T))$, taken over all spanning trees T of G. Let $\beta(G) = e - v + 1$ be the first Betti number of G.

<u>Theorem 2.10.</u> (Xuong [39]). The maximum genus of a connected graph G is given by $\gamma_M(G) = \tfrac{1}{2}(\beta(G) - \xi(G))$. $\quad \square$

We give the explicit formula only for $G = K_n$.

<u>Theorem 2.11.</u> (Nordhaus, Stewart and White [20]). The maximum genus of K_n is

$$\gamma_M(K_n) = [\tfrac{1}{4}(n-1)(n-2)]. \quad \square$$

The original proof used the generalized Euler equation $v - e + f = 2 - 2k$ to obtain the formula as an upper bound, and then attained this bound by constructing an appropriate ρ for each value of n. Now Theorem 2.11 is a trivial corollary of Theorem 2.10 (take $T = K_{1,n-1}$).

Duke [11] established an intermediate value theorem for graph imbeddings that leads directly to a characterization theorem for the map imbedding range for connected graphs:

<u>Theorem 2.12</u>. For G a connected graph and k an integer, there exists a map $i: G \to S_k$ if and only if $\gamma(G) \leq k \leq \gamma_M(G)$. □

Combining Theorems 2.8, 2.11, and 2.12, we find:

<u>Corollary 2.13</u>. For $n \geq 3$, there exists a map $i: K_n \to S_k$ if and only if

$$\{\tfrac{1}{12}(n-3)(n-4)\} \leq k \leq [\tfrac{1}{4}(n-1)(n-2)]. \quad \Box$$

For maps in the interior of the imbedding range (that is, those of genus not necessarily equal to either $\gamma(G)$ or $\gamma_M(G)$), those attracting the most attention have been the self-dual and the symmetrical maps.

The <u>dual</u> of an imbedding $i: G \to S_k$ is a pseudograph (loops and multiple edges being allowed) G* having V(G*) equal to the set of regions for i, with two regions being adjacent if they share an edge of G in their boundaries. If G and G* are isomorphic, the imbedding i (or the corresponding map M) is said to be <u>self-dual</u>; in this case G* is actually a graph, since we are assuming G to be a graph. For example, the map of Figure 1 is self-dual. Self-dual imbeddings have been studied extensively, and we cite a sample of results here.

<u>Theorem 2.14</u>   (Heffter [18], Biggs [3], White [33], Pengelley [23], Stahl [30], and Bouchet [6]). The complete graph K_n has a self-dual imbedding if and only if $n \equiv 0$ or $1 \pmod 4$. □

<u>Theorem 2.15</u>   (White [33]). For $w \geq 1$, there exists a self-dual imbedding of some graph G of order n on $S_{n(w-1)+1}$ if and only if $n \geq 4w + 1$. □

111

Theorem 2.16 (Stahl [30]). *The finitely-generated abelian group Γ has a self-dual imbedding (for some $G_\Delta(\Gamma)$) if and only if $\Gamma \neq Z_2$ or Z_3.* □

We now begin our consideration of symmetrical maps. An <u>automorphism</u> of a map $M = (G,\rho)$ is a permutation $\alpha : V(G) \to V(G)$ preserving oriented region boundaries; that is, if $(v_1, v_2,\ldots, v_n)$ is a region for $i : G \to S_k$ corresponding to (G,ρ), then so is $(\alpha(v_1), \alpha(v_2),\ldots, \alpha(v_n))$. The set of all such α, under composition, forms the <u>map automorphism group</u> Aut M, which is clearly a subgroup of Aut G. Equivalently, a map automorphism is a graph automorphism preserving ρ; that is, $\rho_{\alpha(v)} = \alpha \rho_v \alpha^{-1}$, for all $v \in V(G)$. A basic result is the following:

Theorem 2.17 (Biggs [2]). *If $\alpha \in$ Aut M fixes two adjacent vertices, then α must be the identity permutation on $V(G)$.* □

Now if $S = \{(u,v) : [u,v] \in E(G)\}$ denotes the set of <u>sides</u> of G, then the action of Aut M on $V(G)$ induces an action on S, by $\alpha(u,v) = (\alpha(u), \alpha(v))$; then by Theorem 2.17, $|\text{Aut M}| \leq |S|$. Moreover, if (Aut M, S) is a transitive permutation group, then $|\text{Aut M}| = |S|$ and (Aut M, S) is in fact a <u>regular permutation group</u> (that is, a transitive permutation group having its order equal to its degree, the cardinality of the object set); in this case, the map M is also said to be <u>regular</u>. If M is regular, then Aut M is transitive on the vertices and edges of G and on the regions of M. (Note that $M = (G,\rho)$ is regular if and only if $|\text{Aut M}| = 2|E(G)|$.) Regular maps are also called <u>symmetrical maps</u> (see Biggs [4] and Biggs and White [5]), and we adopt this terminology here. For Cayley maps, there is a sufficient condition which is often quite simple to apply.

Theorem 2.18 (Biggs [4]). *If $M = M(\Gamma, \Delta, r)$ is a Cayley map, and if α is a group automorphism for Γ such that $\alpha|_\Delta = r$, then M is a symmetrical map.* □

We note that the permutation $r : \Delta \to \Delta$ of a Cayley map induces a permutation $r' : S \to S$ on the set of sides of the corresponding Cayley graph, by $r'(a,b) = (b, br(b^{-1}a))$; the orbits of r' determine the region boundaries for the imbedding. It is sometimes helpful to regard r' as another permutation (not necessarily cyclic) of Δ (see Biggs [4]): $r'(\delta) = r(\delta^{-1})$.

A map $M = (G,\rho)$ is said to be <u>reflexible</u> if there is an $\alpha \in$ Aut G such that $\rho_{\alpha(v)} = \alpha \rho_v^{-1} \alpha^{-1}$, for all $v \in V(G)$. (Equivalently, $(v_1, v_2, \ldots, v_n)$ is a region if and only if $(\alpha(v_n), \ldots, \alpha(v_2), \alpha(v_1))$ is a region.) A map which is symmetrical and reflexible has an extended automorphism group of order $4|E(G)|$. If a map is symmetrical but not reflexible, we follow Wilson [38] and say that it is <u>chiral</u>. Symmetrical maps have been studied quite extensively; see, for example, Coxeter and Moser [10]; we also list:

<u>Theorem 2.19</u>. The map $i : G \to S_0$ is symmetrical if and only if $G = C_n$ (an n-cycle, with $n \geq 2$), or the 1-skeleton of a Platonic solid. □

For example, the graph G of Figure 1 is the 1-skeleton of the tetrahedron.

<u>Theorem 2.20</u>. If $i : G \to S_1$ is a symmetrical map, with k denoting the common vertex degree and ℓ the common region size, then $(k,\ell) = (3,6)$, $(4,4)$ or $(6,3)$. □

<u>Theorem 2.21</u> (Brahana [7]). The symmetric groups $S_n (n \geq 3)$ and the alternating groups $A_n (n \geq 4)$ each occur as the automorphism group of a symmetrical map. □

<u>Theorem 2.22</u> (Biggs [2]). The complete graph K_n has a symmetrical map if and only if $n = p^r$ is a prime power. □

For example, the complete maps of Figures 1 and 2 are both symmetrical; the first is reflexible, while the second is chiral. Biggs also showed that any symmetrical map for K_n is in fact a Cayley map for some group of order n (compare with Theorem 6.1), and moreover that Aut M is a <u>Frobenius group</u> (that is, a transitive permutation group for which only the identity permutation can fix two or more distinct objects); this latter observation is an immediate consequence of Theorem 2.17.

The next two results are in contrast to Theorem 2.22:

Theorem 2.23 (Biggs and White [5, §5.67]; see also Stahl and White [31]). The complete bipartite graphs $K_{n,n}$ have symmetrical maps for all n; moreover, these maps can be taken to be 'totally hamiltonian' (that is, each region boundary is a hamiltonian cycle). □

Theorem 2.24 (Biggs and White [5, §§5.66,5.81]). The n-cube Q_n has symmetrical maps for all n, both of genus $\gamma(Q_n)$ and of genus $\gamma(Q_{n+1})$. □

We now attempt to tie together the self-complementation, self-duality, and symmetricality properties. (Note that complementation is a lower-dimensional analog of duality: in taking a dual, 1-dimensional subsets are fixed, while 2- and 0-dimensional subsets are interchanged; in taking a complement, 0-dimensional subsets are fixed, while 1-dimensional subsets are interchanged with '(-1)-dimensional' subsets (the 'non-edges').)

Two permutation groups (G,X) and (G',X') are said to be <u>equivalent</u> if there exist a bijection $\beta : X \to X'$ for the object sets, and a group isomorphism $\phi : G \to G'$ satisfying, for each $x \in X$ and $g \in G$, $\phi(g)(\beta(x)) = \beta(g(x))$; that is, $\phi(g)\beta = \beta g$, so that $\phi(g) = \beta g \beta^{-1}$ and ϕ is induced by β, which may be regarded as a relabelling. For example, (Aut G, V(G)) and (Aut $\bar{G}$, V($\bar{G}$)) are equivalent, with β as the identity permutation. As a less trivial example, if $M = (G,\rho)$ is any map, and $M^* = (G^*, \rho^*)$ its dual, then the actions of Aut M on S and Aut M* on S* are equivalent under $\beta : S \to S^*$, assigning to each $s \in S$ the unique $s^* \in S^*$ 'crossing' s. However, the permutation groups (Aut M, V(G)) and (Aut M*, V(G*)) need not be equivalent; in fact, it is quite possible that $|V(G)| \neq |V(G^*)|$.

The properties for graphs and maps which we have been discussing are symmetry properties; in the self-complementation or self-duality case, the symmetry is external (G compares with $\bar{G}$ or G*, respectively), while in the context of symmetrical maps, the symmetry is internal ($M = (G,\rho)$ compares with itself). We now combine these various symmetries as follows:

<u>Definition</u>. The map $M = (G,\rho)$ is said to be <u>strongly symmetric</u> if (i) $\bar{G} = G$; (ii) $G^* = G$; (iii) M is symmetrical; (iv) Aut M and Aut M* are equivalent under an anti-isomorphism $\beta: V(G) \to V(G^*)$.

3. THE MAIN THEOREM

We now characterize strongly symmetric maps, as to order. We shall need one preliminary result:

Theorem 3.1 (Jordan; see Burnside [8, p.172]). If (G,X) is a Frobenius group of degree $n = |X| \geq 6$ and order $\frac{1}{2}n(n-1) = |G|$, then n is a prime power. □

Theorem 3.2. There exists a strongly symmetric map of order n if and only if n is a prime power congruent to 1 (mod 8).

Proof. Let $M = (G,\rho)$ be a strongly symmetric map, with $n = |V(G)|$ and $e = |E(G)|$. Since G is self-complementary, $e = \frac{1}{2}\binom{n}{2} = \frac{1}{4}n(n-1)$, and $n \equiv 0$ or 1 (mod 4). But since M is symmetrical, G is vertex-transitive, and thus $n \equiv 1$ (mod 4). Now let M imbed G on S_k, with f regions. Since M is self-dual, $f = n$, and the Euler equation $n - e + f = 2 - 2k$ gives $2n - \frac{1}{4}n(n-1) = 2 - 2k$, so that $n \equiv 1$ (mod 8). Since M is symmetrical, $|\text{Aut } M| = 2|E(G)| = \frac{1}{2}n(n-1)$, and Aut M is a transitive permutation group of order $\frac{1}{2}n(n-1)$ and degree $n \geq 9$. We show that Aut M is a Frobenius group and apply Theorem 3.1, to see that n is in fact a prime power.

So, suppose $\alpha \in \text{Aut } M$, u and v are in $V(G)$, with $\alpha(u) = u$, $\alpha(v) = v$, and $u \neq v$. If $[u,v] \in E(G)$, then α is the identity permutation, by Theorem 2.17. If $[u,v] \notin E(G)$, then we apply the anti-isomorphism β giving the equivalence between Aut M and Aut M*. Then β induces an isomorphism ϕ between Aut M and Aut M*, so that $\phi(\alpha) = \beta\alpha\beta^{-1}$. Thus $(\phi(\alpha))(\beta(u)) = \beta\alpha\beta^{-1}(\beta(u)) = \beta\alpha(u) = \beta(u)$, and similarly $\phi(\alpha)$ also fixes $\beta(v)$. But since β is an anti-isomorphism and $[u,v] \notin E(G)$, $[\beta(u), \beta(v)] \in E(G^*)$. Hence, by Theorem 2.17 again, $\phi(\alpha)$ is the identity in Aut M*; but since ϕ is an isomorphism, α is the identity in Aut M.

For the converse, let $n = p^r \equiv 1$ (mod 8), p a prime. We construct a Cayley graph $G_n = G_{\Delta_n}(\Gamma_n)$, where $\Gamma_n = (Z_p)^r$ is the additive group in the Galois field $GF(p^r)$. Let x be a primitive element for $GF(p^r)$, so that the multiplicative group is generated by x; then we take $\Delta_n = \{1, x^2, x^4, \ldots, x^{n-3}\}$, the set of all <u>squares</u> in $GF(p^r)$. (Equivalently, $[u,v] \in E(G_n)$ if and only if $v - u$ is a square in $GF(n)$.) The resulting Cayley graph G_n is called a <u>Paley graph</u> (see [21], where the ideas behind this construction are introduced). We remark that Paley graphs are defined for all prime powers $p^r \equiv 1 \pmod 4$ — since these are precisely the cases for which -1 is a square, so that undirected edges are well defined — but only in the case $p^r \equiv 1 \pmod 8$ are self-dual imbeddings possible.

We now define $r_n : \Delta_n \to \Delta_n$ by $r_n(\delta) = x^2 \delta$, so that $M_n = M_n(\Gamma_n, \Delta_n, r_n)$ is a Cayley map (which we now call a <u>Paley map</u>): r_n induces vertex rotations

$$\rho_v(w) = x^2(w - v) + v,$$

and the permutation $r'_n : \Delta_n \to \Delta_n$, given by

$$r'_n(\delta) = -x^2 \delta.$$

We show that M_n is a strongly symmetric map. (Note that Paley maps are also defined for all $n = p^r \equiv 1 \pmod 4$, but we continue to specialize to $n \equiv 1 \pmod 8$.)

(i) The permutation $\beta : (Z_p)^r \to (Z_p)^r$ given by $\beta(v) = xv$ is an anti-automorphism for G_n, so that G_n is self-complementary. We observe that Paley graphs, being Cayley graphs of odd order, not only have a 2-factor (as required by Theorem 2.4) but in fact are <u>2-factorable</u> (that is, the edge-set can be partitioned so that each subset induces a 2-factor). We also note that the anti-automorphism β has exactly one fixed point (the vertex 0) and one orbit of length $n - 1 \equiv 0 \pmod 4$, consistent with Theorem 2.2.

(ii) To show that (G_n, ρ) is a self-dual map, we study the corresponding imbedding $i : G_n \to S_k$, where $n = p^r = 8m + 1$ and $k = 8m^2 - 7m$ ($m = 1, 2, 3, 5, 6, 9, \ldots$). This imbedding is an $(8m + 1)$-fold covering space of a voltage graph imbedding (see [35], for example) which is the normal form for $S_m : a_1 a_2 \cdots a_{2m} a_1^{-1} a_2^{-1} \cdots a_{2m}^{-1}$. (Note that the values $n = 8m + 1$, $w = m$ are consistent with Theorem 2.15, and that Theorem 2.16 is also illustrated by this construction.) The $2m$ directed edges bounding this $4m$-gon are labelled with generators from Δ_n by the assignment $a_i \to x^{(i-1)(4m-2)}$, $1 \leq i \leq 2m$. This assignment describes, via the theory of voltage graphs, each region boundary (and vertex rotation) in the covering space, the map (G_n, ρ); in particular, the single $4m$-gon below lifts to $8m + 1$ $4m$-gons above. (In Figure 3 we depict the entire situation for $m = 1$, using

$x^2 = 2x + 1$ in GF(9); in Figure 4 we show the voltage graph for m = 2.)

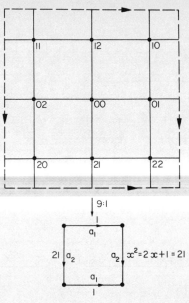

Figure 3   Figure 4

We now employ a method first used by Bouchet [6] for showing self-duality. Each region in the covering space imbedding has a unique lift of the directed edge a_1 from the normal-form voltage graph in its boundary; this lift will be a side (g, g+1) in G_n, where g is uniquely determined for that region; label the region with g*. It is now straightforward to check that, for each $g \in (Z_p)^r$, the region g* (g* $\in V(G_n^*)$) has neighbours N(g*) = $\{(g + \omega(-x^2)^k)^* : 0 \leq k \leq 4m-1\}$ in G_n^*, where $\omega = (1-x^2)/(1+x^2)$. (In fact, the dual is a Cayley map $M_n^*(\Gamma_n, \omega\Delta_n, r_n^*)$, where $r_n^*(\omega\delta) = -x^2\omega\delta$ ($\delta \in \Delta_n$), r_n^* being taken in the opposite sense to that of r_n.) Thus if ω is a square in $GF(p^r)$, then $G_n^* = G_n$; if on the other hand ω is a non-square, then $G_n^* = \bar{G}_n = G_n$.

(iii) Since the permutation r_n of Δ_n extends to a group automorphism of $\Gamma_n = (Z_p)^r$ (this is readily shown, using the distributive law in GF(n)), Theorem 2.18 applies to show that $M_n(\Gamma_n, \Delta_n, r_n)$ is a symmetrical map. In fact, Aut M is a Frobenius group whose Frobenius kernel is the regular normal subgroup $\{\gamma_g : \Gamma_n \to \Gamma_n, \gamma_g(h) = g + h \mid g \in \Gamma_n = V(G_n)\} \cong \Gamma_n$, and whose Frobenius complement is the cyclic group stabilizing the vertex 0, generated by the group automorphism s_0 extending r_n, $s_0(h) = x^2 h$. The other vertex stabilizers are conjugate, with $(\text{Aut } M)_g$ generated by $s_g = \gamma_g s_0 \gamma_g^{-1}$. We note that, of course $|\text{Aut } M_n| = |\Gamma_n||\Delta_n| = 2|E(G_n)|$. Finally, we remark that Aut M is also seen to contain a subgroup isomorphic to Γ_n, by [5, §5.61]; this subgroup must be proper, by Theorem 2.5.

(iv) If $\omega = (1-x^2)/(1+x^2)$ is a non-square, then the anti-isomorphism $\beta : V(G_n) \to V(G_n^*)$ given by $\beta(g) = g^*$ gives an equivalence between Aut M and Aut M*, since if $\alpha^* = \beta\alpha\beta^{-1}$, where α is one of the generating automorphisms s_0 or γ_g ($g \in \Gamma_n$) of Aut M, one readily checks that α^* preserves oriented region boundaries in M*. If, on the other hand, $\omega \in \Delta_n$, then we take $\beta(g) = (xg)^*$ to see again that Aut M and Aut M* are equivalent under an anti-isomorphism β. □

4. COMMENTS ON THE CONSTRUCTION

The Paley maps $M_n = (\Gamma_n, \Delta_n, r_n)$ just constructed have additional properties of interest. To discuss these we need the following facts about Galois fields, taken from Storer [32].

Theorem 4.1. Let x be a primitive element for $GF(p^r)$, where $p^r = 2f + 1$ and f is even. Let (i,j) denote the number of ordered pairs (s,t) such that $x^{2s+i} + 1 = x^{2t+j}$, $0 \le s,t \le f-1$; then

(a) $(0,0) = \frac{1}{2}(f-2)$,

(b) $(0,1) = (1,0) = (1,1) = \frac{1}{2}f$. □

We also need two further definitions. A regular graph which is neither complete nor empty is said to be <u>strongly regular</u> if: (i) the number p_{22}^1 of vertices mutually adjacent with two non-adjacent vertices u and v is independent of u and v, and (ii) the number p_{22}^2 of vertices mutually adjacent with two adjacent vertices u and v is independent of u and v. A $(v, b, r, k; \lambda_1, \lambda_2)$ <u>partially balanced incomplete block design</u> (PBIBD) is an arrangement of v objects into b blocks, with two distinct objects being first or second associates, as determined by non-adjacency or adjacency respectively in a strongly regular graph, so that: (i) each object appears in exactly r blocks, (ii) each block contains exactly k objects, and (iii) each pair of i-th associates (i = 1,2) appear together in exactly λ_i blocks. (If the associated graph is complete, we drop λ_1, write λ for

λ_2, and obtain a (v, b, r, k, λ) <u>balanced incomplete block design</u> (BIBD).)
The connection between block designs and graph imbeddings is studied in [34],
where most of the examples produced have $\lambda_1 = 0$ and λ_2 (or λ) = 1 or 2.
The Paley maps admit an unbounded selection of λ values.

First we check that the Paley graph G_n is strongly regular (here we allow $n = p^r \equiv 1 \pmod 4$). Let $n = 4m + 1 = 2f + 1$, so that $f = 2m$. From Theorem 4.1(a) we find that $(0,0) = \frac{1}{2}(f-2) = m - 1$. Now clearly $x^{2s} + 1 = x^{2t}$ if and only if $x^{2s+2k} = x^{2t+2k}$; it follows that $p_{22}^2 = m - 1$ for G_n. Similarly, we use Theorem 4.1(b) to see that $p_{22}^1 = (1,1) = \frac{1}{2}f = m$. In particular, we have the following:

<u>Theorem 4.2.</u> For $n = 8m + 1$, the Paley graph G_n is strongly regular, with parameters $p_{22}^1 = 2m$ and $p_{22}^2 = 2m - 1$. □

This theorem overlaps with the next result, which we shall invoke later in this section.

<u>Theorem 4.3.</u> If G is self-complementary and has a symmetrical map, then G is strongly regular.
<u>Proof.</u> In fact, for any graph G having a symmetrical map, since Aut M is transitive on edges and Aut M ≤ Aut G, p_{22}^2 is well defined. But if G is self-complementary, then Aut G is transitive on non-edges as well, so that p_{22}^1 is also well defined. □

Thus the Paley graphs G_n, for $n = p^r \equiv 1 \pmod 8$, are certainly candidates for determining association classes for PBIBDs, and in fact the Paley maps give such designs. The vertices will be the objects of a given design, and vertices appearing in a common region boundary will form the blocks.

<u>Theorem 4.4.</u> The Paley map M_n, for n a prime power $= 8m + 1$, yields an $(8m+1, 8m+1, 4m, 4m; \lambda_1, \lambda_2)$ - PBIBD, where $\lambda_1 + \lambda_2 = 4m - 1$, and $\lambda_1 = 2m$ if $x^2 + 1$ is a square in GF(n) and $\lambda_1 = 2m - 1$ if $x^2 + 1$ is a non-square.
<u>Proof.</u> It is notationally convenient to select the region we have labelled ω^*, $\omega = (1-x^2)/(1+x^2)$, which has counter-clockwise boundary determined by:
$$0, 1, 1-x^2, 1-x^2+x^4, \ldots, 1-x^2+x^4- \ldots +x^{8m-4}.$$

It will suffice to examine the differences inherent in this set, as (by the voltage graph theory) this region generates all others. (Region $(\omega+g)*$ is given by $g, g+1, g+1-x^2, g+1-x^2+x^4, \ldots$, for $g \in (Z_p)^r$, and thus will produce exactly the same differences.) Form the $4m \times 4m$ matrix $D = (d_{ij})$, where $v_1 = 0$ and

$$v_k = \sum_{\ell=2}^{k} (-1)^\ell x^{2\ell-4}, \text{ for } 2 \leq k \leq 4m,$$

and $d_{ij} = v_j - v_i$. For $h = 1, 2, \ldots, 4m-1$, let

$$D_h = \{d_{ij}: j - i = h \text{ in } Z_{4m}\}.$$

Then

$$D_1 = \{1, -x^2, x^4, -x^6, \ldots, x^{8m-4}, -x^{8m-2}\},$$
$$D_2 = \{1-x^2, -x^2+x^4, x^4-x^6, \ldots, -x^{8m-2}+1\} = (1-x^2)D_1,$$
$$D_3 = (1-x^2+x^4)D_1,$$
$$\vdots$$
$$D_h = (1-x^2+x^4 \ldots +(-1)^{h+1}x^{2h-2})D_1 = \frac{1 + (-1)^{h+1}x^{2h}}{1 + x^2} D_1, \ 1 \leq h \leq 4m-1.$$

Thus each difference subset consists either of all the squares or of all the non-squares in $GF(n)$. We need only determine how often each situation occurs. But the set $\{(-1)^{h+1}x^{2h}: 1 \leq h \leq 4m-1\}$ consists of every square except -1, and since we know that the number of ordered pairs (s,t) such that $1 + x^{2s} = x^{2t}$, $0 \leq s, t \leq 4m-1$, is given by $(0,0) = \frac{1}{2}(f-2) = 2m - 1$ (by Theorem 4.1(a)), and since $1 + (-1) = 0$ is neither a square nor a non-square, we conclude that exactly $2m - 1$ of the numerators $1 + (-1)^{h+1}x^{2h}$ are squares. Thus if $x^2 + 1$ is a square, then $\lambda_2 = 2m - 1$ and $\lambda_1 = 2m$, whereas if $x + 1$ is a non-square, then $\lambda_1 = 2m - 1$ and $\lambda_2 = 2m$. □

It can be shown that in a Paley map, either each region boundary consists of neighbours $N(v)$, $v \in (Z_p)^r$, or each region boundary consists of neighbours in the complementary graph, so that the designs constructed above also exist in a natural context apart from the topological one given. The topological context does, however, facilitate the next observation, which is that these designs are self-dual, in a very strong sense.

Theorem 4.5. Given a Paley map, the dual of the design of the map and the design of the dual of the map coincide, and both are isomorphic to the design of the Paley map itself.

Proof. The design of the dual has as its objects the blocks (regions) of the original map, and a block in the design of the dual corresponds to an object (vertex) in the original map and contains all the dual objects which represent blocks to which the object belongs – this is precisely the definition of the dual of the design of the original map. Next, we find that the blocks of the map are

$$B_g = \{g + \frac{1 + (-1)^{k+1}x^{2k}}{1 + x^2} : 0 \le k \le 4m-1\}, \quad g \in (Z_p)^r,$$

while in the dual we have

$$B_g^* = \{(g + \frac{1 - x^{2k}}{1 + x^2})* : 0 \le k \le 4m-1\}, \quad g \in (Z_p)^r.$$

Since both sets $\{(-1)^{k+1}x^{2k}: 0 \le k \le 4m-1\}$ and $\{-x^{2k}: 0 \le k \le 4m-1\}$ agree with the set of all squares, the assignment $g \to g*$ gives the desired design isomorphism. □

We make several additional comments in regard to Theorem 3.2. Firstly, we observe that self-dual imbeddings of self-complementary graphs need not be unique for a given order. Moreover, the order need not be a prime power. (We are temporarily **relinquishing** the symmetricality condition for strongly symmetric maps. For the first example, we take $\Omega = \{5; 1, 6, 11, 16, 21\}$ and $\Delta = \Omega \cup \Omega^{-1}$ in $\Gamma = Z_{25}$, so that $G_\Delta(\Gamma)$ is the composition $C_5[C_5]$. Since $\overline{C}_5 = C_5$ and $\overline{G[H]} = \overline{G}[\overline{H}]$ in general, $C_5[C_5]$ is immediately seen to be self-complementary. A self-dual imbedding is constructed by an easy application of Theorem 5.10 of [33]. Since $C_5[C_5]$ is not strongly regular, the map M constructed is not symmetrical, by Theorem 4.3; thus this map differs from the Paley map M_{25} of the same order. In fact, Aut $M \simeq \Gamma = Z_{25}$ and is equivalent to Aut M* (both are Frobenius groups) under the anti-isomorphism $\beta(g) = (2g)*$. (Every map M of a $G_\Delta(\Gamma)$ covering a normal form voltage graph has $\Gamma \simeq \Gamma' \leq$ Aut M; in this case the fact that there are no additional automorphisms follows from the observation that each region boundary contains repeated vertices.) We mention in passing that Theorem 5.10 of [33] also provides a self-dual imbedding, but not a symmetrical map, for the Paley graph G_{25}.

For the second example, we take Ω = {13; 5, 20, 15; 1, 4, 16; 3, 12, -17; 7, 28, -18; 11, -21, -19} in $\Gamma = Z_{65}$; $\theta(g) = 2g$ gives an anti-automorphism, so that $G_\Delta(\Gamma)$ is self-complementary. Again, a self-dual imbedding is constructed, using Theorem 5.10 of [33], but again this map fails to be regular, by Theorem 4.3. Again, Aut M and Aut M* are equivalent under the anti-isomorphism $\beta(g) = (2g)*$, and both are Frobenius groups isomorphic to $\Gamma = Z_{65}$. The non-prime-power order is possible, because the automorphism group is not large enough for the Theorem 3.1 to apply.

We now give another condition sufficient to give prime-power order.

Theorem 4.6. Let G be strongly regular (with $p^1_{22} = 2m$, $p^2_{22} = 2m-1$), and let $M = (G,\rho)$ be symmetrical, yielding an (8m+1, 8m+1, 4m, 4m; 2m-1, 2m) - PBIBD. Then $|V(G)| = 8m + 1$ is a prime power.

Proof. We show that $\alpha \in$ Aut M fixing two distinct vertices u and v must fix all vertices, and apply Theorem 3.1 (since $|$Aut $M| = \frac{1}{2}n(n-1)$, where $|V(G)| = n \geq 9$, the theorem applies). So, let $\alpha(u) = u$, $\alpha(v) = v$. If $[u,v] \in E(G)$, we are done, using Theorem 2.17 again. If $[u,v] \notin E(G)$, then u and v are in exactly $\lambda_1 = 2m - 1$ regions (4m-gons) together, and α must permute these regions. Let $(u,y,\ldots,v,\ldots,x)$ be one such region. If y is one of the $p^1_{22} = 2m$ vertices adjacent to u and to v, then α permutes $2m - 1$ such vertices y, in the $2m - 1$ regions containing u and v (with no regions being fixed, or we are done). Thus α fixes the remaining vertex in the set of vertices adjacent to u and v, so that α is the identity. If, on the other hand, y is one of the $4m - 2m = 2m$ vertices adjacent to u but not to v, then α permutes $2m - 1$ such vertices in these $2m - 1$ regions, and fixes the one remaining; again, α must be the identity. □

Recall from Theorem 4.4 that the Paley map constructed will be as in Theorem 4.6 if and only if $x^2 + 1$ is a non-square. We conjecture that for $p^r \equiv 1 \pmod 8$, one can always find a primitive x in $GF(p^r)$ so that $x^2 + 1$ is a non-square; thus Theorem 4.6 would become a characterization. It is not difficult to verify the conjecture in the special case where $p^r - 1 = 8m$ is a power of 2, as then the number of distinct values of $x^2 + 1$ is $2m$, as x ranges over the set of $4m$ primitives; yet $x^2 + 1$ is a square precisely $2m - 1$ times. In this special case, it is also easy to see that regular maps for Paley graphs are unique for a fixed $p^r - 1 = 2^s$ ($s \geq 3$), in the sense that the genus of the ambient surface is uniquely determined. We comment that the Sylow 2-subgroups of Aut M are precisely the vertex stabilizers.

We remark next that any map M for $G_\Delta(\Gamma)$ covering a normal-form voltage graph as in the Paley map construction will have Aut M and Aut M* equivalent, providing that ρ_0 extends to an automorphism of the group Γ; this is shown by a tedious, but routine, calculation.

Finally, we note that for the Paley graphs G_n ($n = p^r \equiv 1 \pmod 4$), Aut G_n consists not only of the map automorphisms Aut M_n as in (iii) of the proof of Theorem 3.2, but also of the field automorphisms generated by $\theta: (Z_p)^r \to (Z_p)^r$, $\theta(g) = g^p$; in fact, Aut G_n = <Aut M, θ>.

Theorem 4.7 (Carlitz [9]).

Aut $G_n = \{g \to x^{2k} g^{p^s} + a: 0 \leq k \leq 4m-1, 0 \leq s \leq r-1, a \in (Z_p)^r\}$. □

We use this result to study the reflexibility property for Paley maps.

Theorem 4.8. The Paley map M_n is reflexible if and only if $n = 9$. (Thus M_n is chiral if and only if $n \neq 9$.)

Proof. If M_n is reflexible, with $n = p^r$, then we must have $r = 2s$ and

$$\theta^s(g) = g^{p^s},$$

giving a reflection. Since θ^s fixes both 0 and 1, we must have

$$\theta^s(x^2) = x^{2p^s} = x^{p^{2s}-3},$$

123

since $\rho_0 = (1, x^2, \ldots, x^{p^{2s}-3})$. Thus $p^{2s} - 3 = 2p^s$, so that p divides 3; hence $p = 3$ and $s = 1$; that is, $n = 9$.

Conversely, $\theta(g) = g^3$ is readily seen to be a reflection for $n = 9$; refer to Figure 3. □

5. RELATED MAPS

As we have seen, the Paley maps defined for prime powers congruent to 1 (mod 8) have several interesting properties: for example, they are regular self-dual imbeddings of strongly regular self-complementary graphs which produce block designs. In this section we shall see how closely we can approximate these properties by similar maps, for all other prime power orders.

$p^r \equiv 5$ (mod 8). The Paley graphs G_n are defined and are self-complementary and strongly regular, as before. The Euler formula disallows self-dual surface imbeddings, so we utilize pseudosurfaces. Again, we use a normal-form voltage graph, but for $n = 8m + 5$ it is a $(4m + 2)$-gon and thus has two vertices. (See Figure 5 for the case $m = 1$.) The $(8m+5)$-fold covering imbeddings will have $16m + 10$ vertices, each of degree $2m + 1$. If, for each $g \in GF(n)$, the two vertices (a,g) and (b,g) are identified, the result is a symmetrical self-dual pseudosurface imbedding of G_n, for $m > 0$; $G_5^* = C_5^*$ consists of five disjoint loops. (For the pseudosurface theory of voltage graphs, see Garman [13].) In fact the map is a Cayley map $M(\Gamma_n, \Delta_n, r)$, with $r : \Delta_n \to \Delta_n$, $r(\delta) = x^4\delta$, having <u>two</u> cycles. The dual G^* is G if $\omega = (1-x^4)/(1+x^4)$ is a square, or $\bar{G}$ if ω is a non-square. The automorphism group is Frobenius, with Frobenius kernel $\simeq (Z_p)^r$. The stabilizer of vertex 0 is generated by $g \to x^2 g$, which alternates between the two orbits of the vertex rotation at 0. An $(8m+5, 8m+5, 4m+2, 4m+2; 2m, 2m+1)$-PBIBD is determined, if $1 + x^4$ is a non-square; otherwise, the two λ values interchange.

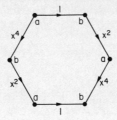

Figure 5

$p^r \equiv 3 \pmod{4}$. Here the Paley graphs are no longer defined, since if $p^r = 4m + 3$, then $-1 = x^{2m+1}$ is not a square. Instead we define the Paley tournament T_n by $(g_1,g_2) \in E(T_n)$ if and only if $g_2 - g_1$ is a square in $GF(n)$; of course, we still take $\Gamma_n = (Z_p)^r$ as our vertex-set. Then $g_1 \neq g_2$ gives $|\{(g_1,g_2),(g_2,g_1)\} \cap E(T_n)| = 1$, so that we do have a tournament. We make the trivial observation that the underlying undirected graph is complete, and hence degenerately strongly regular. The Paley tournaments are self-converse, under the anti-automorphism $\theta: g \to xg$. In Biggs [3] we find symmetrical imbeddings of the associated K_{4m+3} having $8m+6$ regions, each of length $2m + 1$. (These are actually Cayley maps $M(\Gamma_n, \Gamma_n - \{0\}, r)$, where $r(g) = xg$.) Thus the imbeddings cannot be self-dual directly, although they do have bichromatic dual (see [34]) and yield one $(4m+3, 8m+6, 4m+2, 2m+1, 2m)$ - BIBD and two $(4m+3, 4m+3, 2m+1, 2m+1, m)$ - BIBDs - the latter being Hadamard designs. (For example, the two Steiner triple systems arising for $m = 1$ can be read off directly from Figure 2.)

Now we can modify this map to form a self-dual pseudosurface imbedding of K_{4m+3}. Each region contains exactly one edge corresponding to $1 \in (Z_p)^r$, in one of the two possible senses (clockwise or counter-clockwise). In fact, this distinction determines the 2-colouring of the dual. The region is assigned the label g^* ($g \in (Z_p)^r$) if either (i) $(g,g+1)$ bounds the region in the clockwise sense, or (ii) $(g-a, g-a-1)$ bounds the region in the clockwise sense, where $a = 2/(x^{2m}-1)$. Thus each g^* appears exactly twice as a region label, and if these n pairs of vertices in the dual are identified, a self-dual psuedosurface imbedding of K_{4m+3} results.

125

(For example, the two regions labelled 0* for m = 1 are indicated in Figure 2.) Moreover, if the edge directions are carried over into the dual, then we have a self-dual imbedding of the self-converse tournament T_n, with each vertex neighbourhood $N(g^*)$ partitioned into two sets (corresponding to the g^* identification): one consisting of those vertices dominated by g^*, the other consisting of those vertices dominating g^*.

$p^r = 2^r$. Here not even 1 is a square $(1 = x^{2^r-1})$ in $GF(2^r)$, so our previous constructions do not seem to apply. However, if we use the planar voltage graph of Figure 6 with $n = 2^r$, we obtain an n-fold covering imbedding of a 2-fold K_n (each edge appears twice), with $n(n-1)$-gons and $\frac{1}{2}n(n-1)$-digons. If each digon is closed up (by identifying its two edges), a symmetrical self-dual imbedding (see [3]) of the strongly regular K_n (self-complementary in the 2-fold K_n) results. (In fact, $r_n^* = r_n$ for this case.) The concomitant design is an $(n, n, n-1, n-1, n-2)$ - BIBD, which, we note, can be even more readily constructed by taking complements of singletons as the blocks.

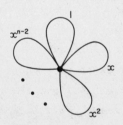

Figure 6

6. UNIQUENESS

Now we explore conditions under which the Paley maps are unique in some sense. Self-duality, strong-regularity, and designs do not enter into the characterizations we present, but as it is Paley maps which we are characterizing, these additional properties persist (except that self-duality fails for $n \equiv 5 \pmod 8$).

Theorem 6.1. If $M = (G,\rho)$ is vertex-transitive, and if (Aut M, V(G)) is self-equivalent under an anti-automorphism θ of G, then M is isomorphic to a Cayley map $M(\Gamma, \Delta, r)$, where $|\Gamma| = |V(G)|$, and $|\Delta| = \frac{1}{2}(|V(G)| - 1)$.

Proof. As in the proof of Theorem 3.2, we find that Aut M is a Frobenius group, so that (see [5], for example) Aut M contains a regular normal subgroup Γ, $|\Gamma| = |V(G)|$. Let $V(G) = \{0,1,\ldots,n-1\}$, and let $\beta: V(G) \to \Gamma$ be given by $\beta(i) = \beta_i$, where uniquely $\beta_i(0) = i$, $0 \le i \le n-1$. Now form $\Delta = \{\beta_i: \beta_i(0) \in N(0)\}$. Since G is vertex-transitive and self-complementary, $|\Delta| = \frac{1}{2}(|V(G)| - 1)$. Since $\beta_0(0) = 0 \notin N(0)$, β_0 (the identity of Γ) $\notin \Delta$. Suppose $\beta_i \in \Delta$, so that $\beta_i(0) = i \in N(0)$; then $\beta_i^{-1}(i) = 0$, and $\beta_i^{-1} \in \text{Aut}(M) \le \text{Aut}(G)$, so that $\beta_i^{-1}(0) \in N(0)$. Thus $\beta_i^{-1} \in \Delta$, and Δ is a generating set for Γ, so that we have a Cayley graph $G_\Delta(\Gamma)$. We claim that G and $G_\Delta(\Gamma)$ are isomorphic, with β giving the isomorphism. So let $[i,j] \in E(G)$. Since $\beta_i^{-1} \in \text{Aut}(M) \le \text{Aut}(G)$, and $\beta_k(0) = k$ for each $k \in V(G)$, we find that $[\beta_i^{-1}(i), \beta_i^{-1}(j)] = [0, \beta_i^{-1}\beta_j(0)] \in E(G)$; that is, $\beta_i^{-1}\beta_j(0) \in N(0)$, and $\beta_i^{-1}\beta_j \in \Delta$. Thus $[\beta(i), \beta(j)] = [\beta_i, \beta_j] \in E(G_\Delta(\Gamma))$, and β is a graph isomorphism.

Now define $r(\beta_i) = \beta_{\rho_0(i)}$, $i \in N(0)$, a cyclic permutation of Δ. We need only show that the maps (G,ρ) and $M(\Gamma, \Delta, r)$ are isomorphic. Let r induce σ by $\sigma_{\beta_i}(\beta_j) = \beta_i r(\beta_i^{-1}\beta_j)$; then $M(\Gamma, \Delta, r) = (G_\Delta(\Gamma), \sigma)$, and we complete the proof by observing that the argument of Biggs [4] applies directly, to give $\sigma = \beta\rho\beta^{-1}$. □

Theorem 6.2. There exists a symmetrical imbedding of a self-complementary graph G of order n, with (Aut M, V(G)) self-equivalent under an anti-automorphism θ of G, if and only if n is a prime power congruent to 1 (mod 4). Moreover, if $\theta^2 \in \text{Aut M}$, the maps are essentially unique, with at most six exceptions.

Proof. (i) Since M is symmetrical and G is self-complementary, $n \equiv 1$ (mod 4) and $|\text{Aut M}| = \frac{1}{2}n(n-1)$. As in the proof of Theorem 3.2, we find that Aut M is a Frobenius group, and thus n is a prime power.
(ii) We use the Paley maps of Theorem 3.2 for $n \equiv 1$ (mod 8) and their surface analogs for $n \equiv 5$ (mod 8). (The Paley maps in the latter case are not self-dual, but have $f = 2v$.) The function $\theta: (Z_p)^r \to (Z_p)^r$ given by $\theta(g) = xg$ is an anti-automorphism giving a self-equivalence for $(\text{Aut M}, (Z_p)^r)$ in each case, since if $\alpha \in \text{Aut M}$, then $\theta\alpha\theta^{-1} \in \text{Aut M}$.

(iii) Just as Aut G is a subgroup of index 2 in $\widehat{\text{Aut}}$ G = <Aut G, θ>, so too is Aut M a subgroup of index 2 in $\widehat{\text{Aut}}$ M = <Aut M, θ>, provided that $\widehat{\text{Aut}}$ M = Aut M $\cup$ (Aut M)θ; that is, if Aut M $\cup$ (Aut M)θ is a group. But this is easy to verify, given that $\theta\alpha\theta^{-1} \in$ Aut M for all α, and $\theta^2 \in$ Aut M. Now, since $|\text{Aut M}| = \frac{1}{2}n(n-1)$ and Aut M is transitive on directed edges, and θ is an anti-automorphism, $|\widehat{\text{Aut}} \text{ M}| = n(n-1)$, and $\widehat{\text{Aut}}$ M is sharply 2-transitive on V(G). By a theorem of Zassenhaus (Theorem 20.3 in Passman [22], for example), we must have either (a) V(G) = GF(n) and $\widehat{\text{Aut}}$ M $\leq$ S(n) = $\{w \to a\sigma(w)+b: a,b \in GF(n), a \neq 0, \sigma \text{ a field automorphism}\}$, or (b) $n = p^2$, with $p = 5, 7, 11, 23, 29$ or 59, where V(G) is identified with the elements of one of the seven irregular nearfields. (There are two possibilities for $p = 11$.)

We now claim that if M = (G,ρ) and V(G) = GF(n), $n = p^r = 4m + 1$, and if we disallow the field automorphism σ, then G is a Paley graph and ρ is determined by the square of a primitive. To see this, first consider α extending ρ_0; then $\alpha(w) = x^k w + b$ and $\alpha(0) = 0$ give $b = 0$, so that $\alpha(w) = x^k w$ and α has order 2m. Thus $2m = 4m/(4m,k)$, so that $(4m,k) = 2$ and k is even. Thus x^k is the square of a primitive element in GF(n), and the neighbourhood N(0) consists either of all squares or of all non-squares. But, by Theorem 6.1, this determines all other vertex rotations. Thus either every edge difference is a square and G is a Paley graph, or every edge difference is a non-square and $\overline{G}$ = G is a Paley graph. Thus if $\theta^2 \in$ Aut M, the maps are unique to within field automorphisms and the choice of a primitive element, except possibly when $n = p^2$, $p = 5, 7, 11, 23, 29$ or 59. □

Corollary 6.3. For each prime $p \equiv 1 \pmod 4$, there exists a unique symmetrical imbedding of a self-complementary graph G of order p having Aut M = Aut G.

Proof. The Paley maps give existence. But if Aut M = Aut G, then θ^2 and $\theta\alpha\theta^{-1}$ (for each $\alpha \in$ Aut M, and where θ is an anti-automorphism of G) are both in Aut M, so that Theorem 6.2 and its proof show that M must be a Paley map. □

The strength of the uniqueness claim is illustrated by Theorem 2.3; the Paley maps of orders 13 and 17, for example, are unique among 5,600 and 11,220,000 self-complementary graphs respectively (as being symmetrical with Aut M = Aut G).

<u>Corollary 6.4</u>. If there exists a symmetrical imbedding of a self-complementary graph G of order n, with Aut M = Aut G, then (with at most six exceptions) n is a prime congruent to 1 (mod 4).
<u>Proof</u>. As in the proof of Corollary 6.3, we find by Theorem 6.2 that n is a prime power congruent to 1 (mod 4); moreover, the maps (with at most six exceptions) are unique and are thus Paley maps. But if $n = p^r$ with $r > 1$, then Aut M is a proper subgroup of Aut G; thus $r = 1$ and n is prime. □

We close with a final characterization. A permutation group (H,X) is said to be $\frac{3}{2}$ - <u>transitive</u> if it is transitive and, for $x \in X$, the stabilizer H_x is non-trivial and all orbits of $(H_x, X - \{x\})$ have the same size. The following is Theorem 17.8 in Passman [22]:

<u>Theorem 6.5</u>. Let G be a $\frac{3}{2}$ - transitive permutation group, and let N be a regular normal subgroup. If G is not a Frobenius group, then N is an elementary abelian p-group for some prime p (and N is the unique minimal normal subgroup of G). □

The above result leads directly to:

<u>Theorem 6.6</u>. There exists a symmetric map M for a self-complementary graph G of order n, with Aut M $\frac{3}{2}$ - transitive and having a regular normal subgroup N, if and only if $n = p^r \equiv 1$ (mod 4), where p is a prime.
<u>Proof</u>. Again the Paley maps establish the sufficiency since, for $n = p^r$, $\Gamma_n = (Z_p)^r$ is a regular normal subgroup of Aut M_n and, for each $g \in V(G_n)$, where $G_n = G_\Delta (\Gamma_n)$, $((\text{Aut } M_n)_g, V(G_n) - \{g\})$ has exactly two orbits (the neighbourhoods of g in G_n and in $\bar{G}_n$), both of size $\frac{1}{2}(n-1)$.

For the converse, the proof of Theorem 3.2 applies, first to give $n \equiv 1$ (mod 4), and then, if Aut M is a Frobenius group, $n = p^r$ for some prime p. On the other hand, if Aut M is not Frobenius, then Theorem 6.5 says that the regular normal subgroup N is an elementary abelian p-group;

thus $n = |N| = p^r$. □

7. QUESTIONS

There remain several questions for further study.

(1) Is there a reasonable characterization for self-dual, self-complementary imbeddings of regular graphs, irrespective of conditions on the map automorphism group? A conjecture might be that such imbeddings occur if and only if $|V(G)| \equiv 1 \pmod 8$; we have exhibited one for $|V(G)| = 65$, but the technique used fails for $|V(G)| = 33$.

(2) What if even the graph regularity is no longer assumed?

(3) Can non-orientable analogs of the maps we have been studying be produced?

(4) If M is the map corresponding to a symmetrical imbedding of a self-complementary graph G, can one always find an anti-automorphism θ of G giving (Aut M, V(G)) self-equivalent and such that $\theta^2 \in$ Aut M (as in Theorem 6.2)?

(5) For $GF(p^r)$, where $p^r \equiv 1 \pmod 8$, can one always find a primitive x such that $x^2 + 1$ is a non-square (as conjectured following Theorem 4.6)?

Acknowledgement. The author is indebted to William Kantor for calling his attention to Theorem 4.7 and for conversations relevant to Section 6, to Norman Biggs for stimulating his interest in symmetrical maps, and to Royal Holloway College, where this research was conducted during his sabbatical year.

REFERENCES

1. K. Appel and W. Haken, Every planar map is four colorable, Bull. Amer. Math. Soc. 82 (1976), 711-712; MR 54 - 12561.
2. N. L. Biggs, Automorphisms of imbedded graphs, J. Combinatorial Theory 11 (1971), 132-138; MR 44 - 3921.
3. N. L. Biggs, Classification of complete maps on orientable surfaces, Rend. Mat. VI 4 (1971), 645-655; MR 48 - 140.
4. N. L. Biggs, Cayley maps and symmetrical maps, Proc. Cambridge Phil. Soc. 72 (1972), 381-386; MR 46 - 1626.
5. N. L. Biggs and A. T. White, Permutation Groups and Combinatorial Structures, Cambridge University Press, Cambridge, 1969.
6. A. Bouchet, Immersions autoduales de graphes de Cayley dans les surfaces orientables, C. R. Acad. Sci. Paris (A) 280 (1975), 1669-1672; MR 52 - 5458.

7. H. R. Brahana, Regular maps and their groups, Amer. J. Math. <u>49</u> (1927), 268-284.

8. W. Burnside, Theory of Groups of Finite Order, Dover, New York, 1955; MR <u>16</u> - 1086.

9. L. Carlitz, A theorem on permutations in a finite field, Proc. Amer. Math. Soc. <u>11</u> (1960), 456-459; MR <u>22</u> - 8005.

10. H. S. M. Coxeter and W. O. Moser, Generators and Relations for Discrete Groups, Springer-Verlag, New York, 1965; MR <u>30</u> - 4818.

11. R. A. Duke, The genus, regional number, and Betti number of a graph, Canad. J. Math. <u>18</u> (1966), 817-822; MR <u>33</u> - 4917.

12. J. Edmonds, A combinatorial representation for polyhedral surfaces, Notices Amer. Math. Soc. <u>7</u> (1960), 646.

13. B. L. Garman, Voltage graph imbeddings and the associated block designs, J. Graph Theory <u>3</u> (1979), to appear.

14. R. A. Gibbs, Self-complementary graphs, J. Combinatorial Theory (B) <u>16</u> (1974), 106-123; MR <u>50</u> - 188.

15. J. L. Gross, Voltage graphs, Discrete Math. <u>9</u> (1974), 239-246; MR <u>50</u> - 153.

16. J. L. Gross and T. W. Tucker, Quotients of complete graphs, revisiting the Heawood map-coloring theorem, Pacific J. Math. <u>55</u> (1974), 391-402; MR <u>52</u> - 10466.

17. L. Heffter, Über das Problem der Nachbargebiete, Math. Ann. <u>38</u> (1891), 477-508.

18. L. Heffter, Uber metacycklische Gruppen und Nachbar-configurationen, Math. Ann. <u>50</u> (1898), 261-268.

19. C. K. Lim, On graphical regular representations of direct products of groups, J. Combinatorial Theory (B) <u>24</u> (1978), 242-246.

20. E. A. Nordhaus, B. M. Stewart and A. T. White, On the maximum genus of a graph, J. Combinatorial Theory (B) <u>11</u> (1971), 258-267; MR <u>44</u> - 3922.

21. R. E. A. C. Paley, On orthogonal matrices, J. Math. Phys. <u>12</u> (1933), 311-320.

22. D. Passman, Permutation Groups, Benjamin, New York, 1968; MR <u>38</u> - 5908.

23. D. Pengelley, Self-dual orientable embeddings of K_n, J. Combinatorial Theory (B) <u>18</u> (1975), 46-52; MR <u>51</u> - 5357.

24. S. B. Rao, Characterization of self-complementary graphs with 2-factors, Discrete Math. <u>17</u> (1977), 225-233.

25. R. C. Read, On the number of self-complementary graphs and digraphs, J. London Math. Soc. <u>38</u> (1963), 99-104; MR <u>26</u> - 4339.

26. G. Ringel, Selbstkomplementäre Graphen, Archiv. der Math. <u>14</u> (1963), 354-358; MR <u>27</u> - 4222.

27. G. Ringel, Map Color Theorem, Springer-Verlag, Berlin, 1974; MR 50 - 1955.

28. G. Ringel and J. W. T. Youngs, Solution of the Heawood map-coloring problem, Proc. Nat. Acad. Sci., U.S.A. 60 (1968), 438-445; MR 37 - 3959.

29. H. Sachs, Über selbstkomplementäre Graphen, Publ. Math. Debrecen 9 (1962), 270-288; MR 27 - 1934.

30. S. Stahl, Self-dual Embeddings of Graphs, Ph.D. Thesis, Western Michigan University, Kalamazoo, 1975.

31. S. Stahl and A. T. White, Genus embeddings for some complete tripartite graphs, Discrete Math. 14 (1976), 279-296; MR 54 - 10060.

32. T. Storer, Cyclotomy and Difference Sets, Markham, Chicago, 1967; MR 36 - 128.

33. A. T. White, Orientable imbeddings of Cayley graphs, Duke Math. J. 41 (1974), 353-371; MR 51 - 230.

34. A. T. White, Block designs and graph imbeddings, J. Combinatorial Theory (B) 25 (1978), 166-183.

35. A. T. White, Graphs of groups on surfaces, Combinatorial Surveys: Proceedings of the Sixth British Combinatorial Conference (ed. P. J. Cameron), Academic Press, London (1977), 165-197.

36. A. T. White, The proof of the Heawood conjecture, Selected Topics in Graph Theory (ed. L. W. Beineke and R. J. Wilson), Academic Press, London, 1978.

37. A. T. White and L. W. Beineke, Topological graph theory, Selected Topics in Graph Theory (ed. L. W. Beineke and R. J. Wilson), Academic Press, London, 1978.

38. S. Wilson, Smallest nontoroidal chiral maps, J. Graph Theory 2 (1978), to appear.

39. N. H. Xuong, How to determine the maximum genus of a graph, J. Combinatorial Theory (B), to appear.

40. J. W. T. Youngs, Minimal imbeddings and the genus of a graph, J. Math. Mech. 12 (1963), 303-315; MR 26 - 3043.

Arthur T. White
Department of Mathematics
Western Michigan University
Kalamazoo
Michigan 49008
USA

C C Wright
Arcs and cars: an approach to road traffic management based on graph theory

1. INTRODUCTION

Most people are familiar with the day-to-day problems of urban traffic congestion, which arise mainly because road networks have limited capacity to carry vehicles between their origins and destinations. In a sense, even the environmental side-effects of traffic such as noise and atmospheric pollution can be attributed to a shortage of network capacity of the right type in the right place. Because road building is expensive and disruptive to the community, the engineering resources of local authorities in towns throughout Europe and America are being channelled more and more into 'traffic management' - that is, low-cost measures which increase the capacity of existing roads or which manipulate traffice in order to satisfy some other objective such as the creation of pedestrian areas. Whether or not the main aim is to increase the overall network capacity, almost every traffic management scheme requires that some of the roads affected by the scheme will be used more intensively in order to cope with re-routed traffic.

The capacity of a road is limited because drivers will not usually tolerate time separations ('headways') between their vehicles which are less than a certain amount - about 1.5 seconds. These headways are most critical at junctions, where drivers arriving from different approaches have to share the same space: to put it another way, the arriving streams interrupt one another, either in a regulated fashion under the control of a policeman or traffic signals, or in a more-or-less random fashion at roundabouts and GIVE WAY junctions. Hence the capacity of a network is largely governed by the capacity of its junctions, just as the average speed attainable on the network depends principally on junction delays.

There are three basic types of interaction or 'conflict' between vehicles at junctions, which are distinguished according to whether the vehicle paths merge, diverge, or cross. Because they involve competition for road space, merging and crossing conflicts generate greater delays and a greater risk of accident than diverging conflicts; nevertheless, they are all important and their numerical frequencies, taken together, give a fair indication of the nature and size of the problem at any given junction. However, in this paper we are more interested in networks as a whole, and for this purpose it is simpler to work in terms of path crossings rather than conflicts, these being defined in the same way as crossings on a graph - that is, as intersections of the arcs corresponding to vehicle paths on the network. The paths of different vehicles passing along the same road are represented as separate arcs spaced out laterally across the width of the road so that 'weaving' movements like the one shown on the right in Figure 1 are considered as crossings in exactly the same way as crossing conflicts at junctions. In this particular case, a single path crossing serves to represent both a merging and a diverging conflict. If, however, the two paths had merged and subsequently diverged without weaving, no path crossing would be deemed to have occurred.

Examples of path crossings

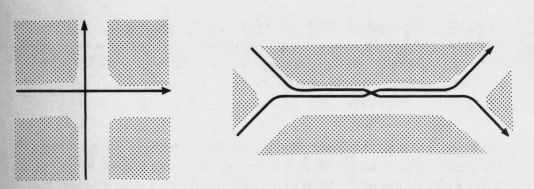

Figure 1

Although not entirely satisfactory for this reason, the rate of occurrence of path crossings over a network per unit time is adopted as a broad measure of the 'efficiency' of the traffic pattern in terms of congestion and accident risk. In fact, traffic management schemes are usually evaluated in terms of the projected time savings to road users or their cash equivalent, together with other money costs and benefits (social and environmental considerations are dealt with separately). However, the estimation of these quantities for any given traffic pattern is a laborious process, so that in practice only a small number of alternative schemes can be quantitatively compared. In urban areas where the number of possible combinations of measures such as turn bans and one-way streets is potentially very large, this means that with existing procedures the 'screening' of alternatives prior to detailed evaluation has to be done intuitively - it is often difficult to see the wood for the trees. Hence a rough and ready measure of performance such as path crossings offers an attractive tool for the preliminary assessment of different options, and may help the planner to come to grips more easily with the fundamental problems of network operation. In particular, one might wish to use such a measure to develop a set of broad general principles or guidelines for manipulating the spatial pattern of traffic, using simple devices such as banned turns and one-way streets, to achieve optimum or near-optimum results. One suspects that the road networks in many towns are not used to the best advantage; while data on this point is scarce, it is believed, for example, that drivers making radial journeys to and from Central London via one particular Inner London Borough cross and re-cross each others' paths at several points on the network, which is clearly unsatisfactory.

The idea of looking at city-wide traffic patterns in this way is not by any means new: Smeed [2] originated the theoretical study of urban networks in terms of geometrical routeing systems, and Holroyd and Miller [1] drew attention to the particular importance of path crossings and calculated their frequency for several different types of idealised network; in either case the networks were simple uniform grid patterns treated as a continuum, with infinitesimal spacing between roads. This paper merely carries the idea a little further, applying it to traffic management problems where the road network must necessarily be treated as a finite collection of individual arcs and vertices.

2. REPRESENTATION OF ROAD TRAFFIC PATTERNS IN TERMS OF GRAPH THEORY

A graphical representation of a road network is not the same thing as a graphical representation of the journeys made on it: the network diagram simply defines the choices available to drivers, while the journey diagram should explicitly represent individual vehicle movements and the interactions between them.

2.1 Road Networks

The network making up the public highway system can be represented as a directed graph with certain distinct features:

(i) On two-way roads in the UK vehicles drive on the left, so that except in the case of one-way streets and flyover junctions the arcs occur in complimentary pairs with a 'clockwise' orientation.

(ii) Vertices can be divided into two classes: firstly, there are terminal vertices, which correspond to the start and finish points of journeys either at the entrances to private buildings, or at parking places, or, since the network under consideration almost always forms part of a wider system, at the points where roads enter and leave the study area at the boundary; and secondly, there are vertices at junctions where alternative routes merge and diverge.

(iii) For clarity, all the possible turning movements at junctions should be explicitly represented by distinct arcs and vertices; the representation of opposing right turns (that is, right turns made from approaches on opposite sides of the junction) is crucial; some junction layouts allow vehicles to pass nearside-to-nearside while others do not, with obvious implications for the frequency of path crossings. Examples of several different junction layouts are shown in the hypothetical network diagram in Figure 2.   A SIMPLE ROAD TRAFFIC NETWORK.

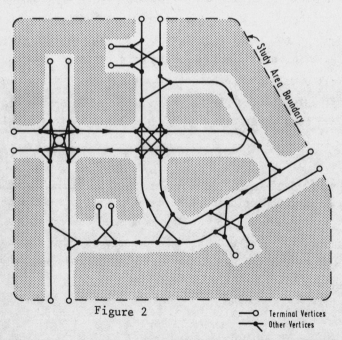

Figure 2

—○ Terminal Vertices
—● Other Vertices

(iv) Road network graphs do not normally contain loops, nor multiple arcs between any two vertices. They are 'strongly connected': a suitably directed chain of arcs can always be found from any origin vertex or junction vertex to any destination or junction vertex. Vertices within junctions are usually only of degree three or four - exceptionally five or more - because most road junctions have only three or four approaches. (The degree may of course be <u>less</u> than the number of approaches, depending on whether all the possible turning movements are permitted.) By way of illustration, Table 1 shows the frequency distribution of the number of approaches at junctions within the 22 km^2 area of the City of Westminster.

Table 1. <u>Frequency distribution of road junctions in Westminster with different numbers of approaches</u>

	Number of approaches*				Total
	3	4	5	6 or more	
Number of junctions	1711	663	19	6	2399
% of all junctions	71%	28%	1%	0$^+$%	100%

* Any road connected to a junction is designated as an 'approach' here, even if it is subject to a one-way restriction which only permits movement away from the junction.

2.2 Vehicle Paths

Distinct from the network itself are the paths actually followed by vehicles - that is, sequences of arcs on the network between two terminal vertices. These correspond more-or-less to what are known as 'k-graphs' in graph theory, the parameter k standing for the number of arcs traversed during any particular journey. These paths have certain characteristic features. In particular, they tend to be 'progressive' in the sense that each successive arc normally takes the driver closer to his destination (more precisely, the remaining distance, measured along the shortest available route from the vehicle's current position to its destination, progressively decreases). The amount of travel distance wasted through non-progressive circulation appears in practice to be quite small. In Central London, for example (Wright [3]), the average distance travelled by vehicles in excess of the shortest-distance route is only about 5%. On the other hand, the number of excess route crossings, measured as a proportion of the number that would occur with shortest-distance routeing, is about 12%, and when measured in relation to the number which would occur with straight-line routeing the excess is as high as 30%.

Measured either way, the excess can largely be accounted for by routeing which is progressive but not optimal - only a very few drivers actually go round in circles, and of those a substantial proportion would probably be looking for parking places.

2.3 Traffic Management as a Graph Theory Problem

For the purpose of this paper we assume that the frequency with which vehicles wish to travel between any two terminal vertices i and j is fixed, at a rate denoted by $q_{i,j}$ vehicles per unit time, and that the pattern of allocation between different alternative routes does not change with time. Hence, the rate at which path crossings occur will be constant. For brevity, each combination of an origin terminal and destination terminal will be referred to as an O-D pair.

We can now state the problem in relatively simple terms: how can the various origin-destination movements be allocated to arcs on the network so as to minimise the total frequency C of path crossings per unit time, subject to reasonable constraints on the distance travelled by each vehicle?

Unfortunately, it is not easy to specify what the travel distance constraints should be, except perhaps in the case of networks based on simple uniform grid patterns (for instance the rectangular grid, where one might stipulate that all routes should be shortest-distance routes). The reason is that path crossings can often be reduced by making drivers follow more tortuous routes, and hence a wide range of strategies may be available in which the total consumption of travel distance can be traded off in varying degrees against the crossing frequency. In order to choose the best strategy one has to have some notion of the 'value' of saving one path crossing as compared with the 'value' of reducing the total travel distance by one unit. Currently, the appropriate rate of exchange is not known.

An alternative and more clear-cut objective would be to ensure as far as possible that no two vehicle paths crossed each other more than once. This would at least help to eliminate unnecessary vehicle interactions, although it would not necessarily minimise them except in the special case where all the journey terminals were on the study area boundary and re-routeing was permitted only within the area enclosed by the boundary.

In the next section we look at some very simple examples of routeing systems on finite networks, and compare their performance with these two alternative objectives in mind. Only in one case is there any possibility of a trade-off between path crossings and journey distance; more complex systems, with more interesting possibilities, remain to be analysed.

3. MINIMISING ROUTE CROSSINGS ON SIMPLE NETWORKS - SOME EXAMPLES

In what follows we shall stipulate that a unit value is assigned to the flow per unit time q_{ij} between each O-D pair i and j ($i \neq j$), so that the problem of minimising the frequency of route crossings is considerably simplified: not only is traffic moving from any particular origin to any particular destination confined to one route - only one vehicle makes the journey - but the total frequency of path crossings per unit time is then simply equal to the number of points at which the various paths intersect. Arbitrary values for the O-D flows would probably lead in general to a mathematical programming formulation.

Note that the terminal vertices on a network diagram can be taken to represent the start and finish points of <u>aggregates</u> of journeys as well as individual ones; in other words they can represent whole <u>areas</u> or zones within a town if desired. However, in the following examples each terminal zone is permitted to have only one connecting link to the network proper, whereas in real life there would usually be several: this restriction is entirely arbitrary, but it does allow one to see more clearly the basic nature of the problem, and to compare different forms of network on an equal basis. Additional crossings would of course occur inside the terminal areas, but they are not taken into account.

3.1 One-to-many and Many-to-one Systems

In the special case of a network having only one origin terminal or one destination terminal, any suitably directed tree will form the basis of a sufficient routeing system which entails no path crossings. Furthermore, in such a case, one or more trees can always be identified which provide shortest-distance routes for every journey. Hence, path crossings and journey distance can be minimised simultaneously. The process of finding shortest-distance trees is in fact an important part of the conventional procedure for forecasting the distribution of traffic volumes over any type of road network when drivers are allowed individual freedom of route choice, and itself constitutes a well-known problem in graph theory.

3.2 Network Equivalent to a Complete Graph with all Terminals on a Convex Boundary

Consider a network consisting of N origins and N destinations arranged alternately as the vertices of a convex polygon, with straight roads connecting each O-D pair as shown in Figure 3. Then it is clear that the total number of path crossings is minimised when all vehicles follow straight routes; the total journey distance is of course minimised simultaneously. If unit traffic flows are assigned to all O-D pairs, it is not difficult to show that the total frequency of path crossings is $\frac{1}{6} N^2 (N - 1)(N - 2)$ per unit time. This value forms a convenient yardstick for comparison with other routeing systems having the same configuration of terminal vertices and the same O-D flows. We shall denote its value by C_0.

STRAIGHT-LINE ROUTEING BETWEEN TERMINALS ON A CONVEX BOUNDARY.

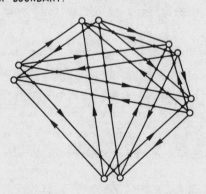

Figure 3

3.3 Corridor Networks

Perhaps the simplest form of network having any practical significance would be a single straight road or 'corridor' with access roads to terminal vertices arranged along one side. An example of the routeing pattern for a two-way corridor connecting five sets of terminals is shown in Figure 4. An alternative arrangement is also shown, with a one-way corridor passing along either side.

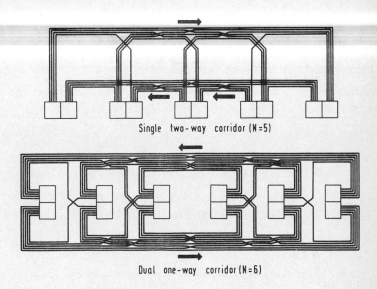

Figure 4

As one might intuitively expect, the dual corridor system entails fewer crossings. Formulae for the frequencies of path crossings are given for each layout in Table 2, together with limiting values for the ratios C/C_0. Note that with either layout only one route exists between each O-D pair, so that there is no possibility of a trade-off between path crossings and journey distance.

Table 2. Path crossings for simple routeing systems with one unit of traffic routed between each origin-destination pair per unit time

Network*	Frequency C of path crossings per unit time (N = No. of origins = No. of destinations)**		$\lim_{N \to \infty} (C/C_0)$
Corridor Systems { Single corridor, two-way	$\frac{1}{4}N(N-1)^2(N-2)$	(N = 3, 4, 5, ...)	1.500
Corridor Systems { Two corridors, one-way	$\frac{1}{12}N(N-1)(N-2)^2$	(N = 4, 6, 8, ...)	0.500
Ring Systems { One ring, one-way	$\frac{1}{6}N(N-1)(N-2)(2N-3)$	(N = 3, 4, 5, ...)	2.000
Ring Systems { One ring Two-way { Equal split	$\frac{1}{24}N(N-1)(5N^2-16N+15)$	(N = 3, 5, 7, ...)	1.250
Ring Systems { One ring Two-way { Optimum split	$\frac{1}{54}N(N-2)(10N^2-13N-5)$	(N = 5, 8, 11, ...)	1.111
Ring Systems { Inner and Outer ring, both one-way, equal split	$\frac{1}{12}N(N-1)(N^2-2N+3)$	(N = 3, 5, 7, ...)	0.500
Direct routeing between vertices of convex polygon (see Figure 3)	$\frac{1}{6}N^2(N-1)(N-2) = C_0$	(N = 3, 4, 5, ...)	—

* For diagrams of the various routeing systems, see Figures 3–5.

** The formulae in this table are valid only for the stated values of N. Formulae for other values, which have been omitted for clarity, follow similar (but not identical) curves, and yield identical limiting values for the ratio C/C_0 in all cases.

3.4 Ring Systems and Roundabouts

Arguably, the next simplest form of network should be a ring system with the terminal vertices arranged either (i) around the outside of the ring, or (ii) around the inside. The first arrangement corresponds in effect to a roundabout system, while the second arrangement might be taken to represent a whole town with a pedestrianised central area and an outer ring road. We shall consider only the first arrangement, the mathematics for the second being identical.

To begin with, we suppose that the ring is clockwise one-way, like a conventional UK roundabout. Again, only one route exists between each O-D pair. A formula for the frequency of path crossings is given in Table 2, and Figure 5(a) shows the configuration of vehicle paths for the case where there are just three pairs of terminals (N = 3). Note that with such a system all the path crossings arise from weaving movements, as opposed to crossing conflicts at junctions; the limiting frequency for large N is considerably higher than that for either of the two corridor systems mentioned earlier.

A reduction in the frequency of crossings can be achieved however by making the ring two-way, as shown in the example with five pairs of terminals in Figure 5(b). Here, each driver has a choice between two routes (clockwise or anti-clockwise). It is a fairly simple but tedious matter to show that if journeys are routed clockwise to the first consecutive destination terminals in that direction, and anti-clockwise to the remaining N - n - 1 destinations, then the total number of the route crossings is minimised if n is set at the nearest integer value to $\frac{1}{3}(N + 1)$. This is perhaps surprising, since it implies that the greater proportion of traffic should go the 'wrong way' round the ring from the engineering point of view. A formula for the frequency of path crossings is given in Table 2 for values of N incremented in steps of three.

The route pattern which minimises path crossings does not in general simultaneously minimise the total journey distance. To minimise total journey distance of course, one simply needs to route each individual journey via the shortest of the two routes available. If the terminals were equally spaced around the ring, this would in effect mean an equal split between the clockwise and anti-clockwise directions of flow. The frequency of crossings in such a pattern is in fact only marginally greater

RING ROUTEING SYSTEMS.

(a) One ring, one way
($N=3$)

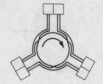

(b) One ring, two way,
traffic split equally
in opposite directions
($N=5$)

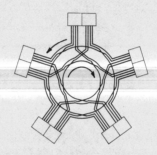

(c) Inner & outer rings,
one way,
traffic split equally
in opposite directions
($N=7$)

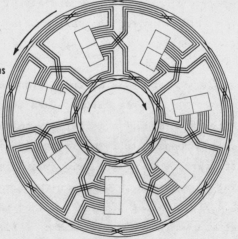

Figure 5

than for the crossing-minimising pattern; a formula is given in Table 2 for odd values of N.

A further reduction in the frequency of path crossings can be obtained by separating the two directions of movement - that is, by placing the anti-clockwise ring around the outside of the terminals as shown in Figure 5(c). A formula is given in Table 2 for the case in which traffic is split equally in opposite directions. This arrangement simultaneously minimises path crossings and total journey distance.

3.5 Comment

The limiting value of the ratio C/C_0 for the dual ring system in Figure 5 turns out to have the same value as for the dual corridor system in Figure 4, and these two systems are the most efficient which the author has been able to find. Although very simple in conception, and analysed here under somewhat artificial assumptions, they are interesting because they give at least a vague general indication of the sort of traffic pattern which might be desirable on efficiency grounds. In principle, either system can be applied to a town with any number of origin and destination zones regardless of their geometrical arrangement, provided they are connected by road links in the required topological configuration.

4. CONCLUSIONS

Seen from the planner's point of view, road traffic management is a fairly complex matter: all that we have been able to do in this paper is to cast the problem in a particular form, rather than provide solutions. Further work seems to be desirable in two main areas. First of all, on the practical side it would be useful to construct an efficient computer-based algorithm for generating alternative routeing patterns for any given network with any given set of O-D flows, which could be used to screen alternative traffic management proposals for urban areas. Ideally, instead of working in terms of path crossings the algorithm would enumerate separately the three basic types of vehicle conflict referred to in Section 1. Secondly, there seems to be a case for more research into the principles of efficient network operation; the study of more complex networks with arbitrary O-D flows will of course require more sophisticated forms of analysis.

One fairly obvious rule of thumb for 'efficient' traffic routeing, applicable to most types of traffic network, is simply that all the vehicles travelling between any particular O-D pair should use the same route; the reason for this is easy to see with the aid of an inductive argument, based on the fact that if some vehicles use a route A which leads them to cross more paths than the vehicles using another route B, the total frequency of crossings would be reduced if all vehicles using A were transferred to B. Another is simply that, other things being equal, left turns are preferable to right turns because they entail fewer conflicts (assuming the left-hand rule of the road).

To summarise, road users in urban areas would probably benefit if the pattern of movement were geometrically tidier, but at the moment no comprehensive rationale is available for manipulating the system except on an intuitive or incremental basis. The aim would be to channel the movement into patterns which minimise conflict, and a theoretical approach along the general lines suggested here may help towards the evolution of some basic principles for generating workable systems.

Acknowledgement

The author is grateful to the City Engineer of Westminster City Council, Mr. A. J. Cryer, for permission to publish this paper, which does not necessarily reflect the views of the City Council.

REFERENCES

1. E. M. Holroyd and A. J. Miller, Route crossings in urban areas, Proc. Australian Road Research Board 3, Pt.1 (1966), 394-419.
2. R. J. Smeed, Road development in urban areas, J. Inst. Highway Engineers 10(1) (1963), 5.
3. C. C. Wright, Some characteristics of drivers' route choice in Westminster, Conference on Urban Transport Planning, University of Leeds, March 1976 (unpublished).
4. C. C. Wright, Control of drivers' route choice: pipe-dream or panacea?, Transportation 7(1978), 193-210.

C. C. Wright
Transport and Safety Research Section
City Engineer's Department
Westminster City Council
London SW1E 6QG
England

M Gordon and J W Kennedy
Poster session paper
Some problems on lattice graphs

The statistics of rubber networks are widely tackled by an unsuitable model, viz. the random insertion of lines to build up a square lattice. The critical point ('Gel Point') of this model features an incipient infinite connected graph ('Gel Molecule') of high cycle rank. In reality the incipient gel molecule is practically tree-like (low cycle rank). Thereafter, further insertion of lines can be modelled on successive random insertion of chords into a spanning tree of a diamond (tetrahedral) lattice graph. This raises the following problems.

Let $\underline{L}$ be an infinite regular (lattice) graph.
Let $T(\underline{L}) = \{t_i : i = 1, \ldots, |T(\underline{L})|\}$ be the set of <u>all</u> spanning trees of $\underline{L}$.
Let $l(t_i)$ be the mean circuit length over all fundamental circuits defined by t_i in $\underline{L}$.
Similarly, let $l(T) = \sum_i l(t_i)/|T(\underline{L})|$ be the mean over all spanning trees. Then:

<u>Problem 1.</u> Is $l(T)$ finite for any lattice graph $\underline{L}$?

<u>Problem 2.</u> If $l(T)$ is not finite, is there a subset $T'(\underline{L}) \subset T(\underline{L})$ for which $l(T')$ is finite? For example, produce a spanning tree t of $\underline{L}$ for which $l(t)$ is finite.

It is not difficult to give an example of a spanning tree h for which $l(h)$ is not finite. Consider $\underline{L}$ = square planar lattice with points in Z^2. Let $\underline{L}_n$ be that (finite) part of it bounded by $0 \leq z \leq n$, so that $\underline{L}_n$ has $(n+1)^2$ points (see Figure 1).

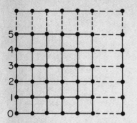

Figure 1

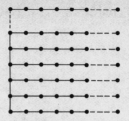

Figure 2

Let h be the spanning tree in Figure 2. Then $T'(\underline{L}_n) = h$ defines n^2 fundamental cycles, comprising n cycles each of length 4, 6, 8, ..., 2(n + 1). Thus

$$l_n(h) = \frac{1}{n^2} \sum_{k=1}^{n} 2(k + 1)n = n + 3.$$

We are unable to construct a spanning tree for which the limit $\lim_{n \to \infty} l_n(h)$ is bounded.

M. Gordon and J. W. Kennedy
Institute of Polymer Science
University of Essex
Colchester CO4 3SQ

QM Library

23 1214991 2

HG 106 BOY

MAIN LIBRARY
QUEEN MARY, UNIVERSITY OF LONDON
Mile End Road, London E1 4NS
DATE DUE FOR RETURN

**WITHDRAWN
FROM STOCK
QMUL LIBRARY**